Protein Biochemistry, Synthesis, Structure and Cellular Functions

PRO-INFLAMMATORY CYTOKINES IN LEARNING AND MEMORY

PROTEIN BIOCHEMISTRY, SYNTHESIS, STRUCTURE AND CELLULAR FUNCTIONS

Molecular Pathology of Proteins
D.I. Zabolotny (Editor)
2009. 978-1-60741-020-1

Serpins and Protein Kinase Inhibitors: Novel Functions, Structural Features and Molecular Mechanisms
Bojidor Georgiev and Sava Markovski (Editors)
2009. 978-1-60741-187-1

Handbook of Lipoprotein Research
Jackson E. Rathbond (Editor)
2010. 978-1-61668-186-9

Proteomics: Methods, Applications and Limitations
Giselle C. Rancourt (Editor)
2010. 978-1-61668-691-8

Protein Engineering: Design, Selection and Applications
Mallorie N. Sheehan (Editor)
2010. 978-1-61668-286-6

Pro-Inflammatory Cytokines in Learning and Memory
Amaicha Mara Depino (Author)
2010. 978-61668-626-0

Protein Engineering: Design, Selection and Applications
Mallorie N. Sheehan (Editor)
2010. 978-1-61668-649-9

Handbook of Lipoprotein Research
Jackson E. Rathbond (Editor)
2010. 978-1-61668-714-4

Proteomics: Methods, Applications and Limitations
Giselle C. Rancourt (Editor)
2010. 978-1-61668-800-4

Pro-Inflammatory Cytokines in Learning and Memory
Amaicha Mara Depino (Author)
2010. 978-1-61668-905-6

Protein Biochemistry, Synthesis, Structure and Cellular Functions

PRO-INFLAMMATORY CYTOKINES IN LEARNING AND MEMORY

AMAICHA MARA DEPINO

Nova Science Publishers, Inc.
New York

Copyright © 2010 by Nova Science Publishers, Inc.

All rights reserved. No part of this book may be reproduced, stored in a retrieval system or transmitted in any form or by any means: electronic, electrostatic, magnetic, tape, mechanical photocopying, recording or otherwise without the written permission of the Publisher.

For permission to use material from this book please contact us:
Telephone 631-231-7269; Fax 631-231-8175
Web Site: http://www.novapublishers.com

NOTICE TO THE READER

The Publisher has taken reasonable care in the preparation of this book, but makes no expressed or implied warranty of any kind and assumes no responsibility for any errors or omissions. No liability is assumed for incidental or consequential damages in connection with or arising out of information contained in this book. The Publisher shall not be liable for any special, consequential, or exemplary damages resulting, in whole or in part, from the readers' use of, or reliance upon, this material.

Independent verification should be sought for any data, advice or recommendations contained in this book. In addition, no responsibility is assumed by the publisher for any injury and/or damage to persons or property arising from any methods, products, instructions, ideas or otherwise contained in this publication.

This publication is designed to provide accurate and authoritative information with regard to the subject matter covered herein. It is sold with the clear understanding that the Publisher is not engaged in rendering legal or any other professional services. If legal or any other expert assistance is required, the services of a competent person should be sought. FROM A DECLARATION OF PARTICIPANTS JOINTLY ADOPTED BY A COMMITTEE OF THE AMERICAN BAR ASSOCIATION AND A COMMITTEE OF PUBLISHERS.

LIBRARY OF CONGRESS CATALOGING-IN-PUBLICATION DATA

Available upon Request
ISBN: 978-1-61668-626-0

Published by Nova Science Publishers, Inc. ✢ *New York*

CONTENTS

Contents		vii
Preface		ix
Chapter 1	Introduction	1
Chapter 2	Sources of Cytokines in the Brain	3
Chapter 3	Cytokines in Disease	7
Chapter 4	Cytokines in Health: Role in Learning and Memory	17
Chapter 5	Conclusion	27
Abbreviations		29
References		31
Index		49

PREFACE

In animals and in humans, cytokines are typically immune system molecules that help orchestrate the host responses to infection. Pro-inflammatory cytokines coordinate immune, physiological, metabolic and behavioral responses that are collectively termed the acute phase reaction. In addition to their role in immunoregulation of inflammatory processes, pro-inflammatory cytokines represent the major communication link between peripheral immunity and the central nervous system. Cytokines are known to play a role in the physiological and behavioral adjustments that occur during sickness, leading to "sickness behavior." One of the most salient symptoms of the sickness behavior syndrome is alteration in learning and memory processes. Studies in animals have demonstrated that acute activation of pro-inflammatory cytokine signaling in the brain in response to peripheral immune activation is associated with deficits in hippocampal-dependent memory.

Recent studies suggest that brain cytokines may also have some physiological roles in neural, neuroendocrine, and behavioral regulation in non-pathological situations. Cytokine-induced modulation of memory processes is a complex phenomenon, including both detrimental and beneficial effects, depending on the specific pro-inflammatory cytokine, its levels (particularly within the brain), and the particular condition that elicits the cytokine secretion. Some cytokines, e.g., Interleukin (IL)-1, have been shown to significantly influence memory consolidation. IL-1, IL-6 and tumor necrosis factor (TNF)-α are the main cytokines involved in learning and memory modulation. These behavioral data are consistent with the impairing effect of enhanced cytokine signaling on hippocampal long-term potentiation.

We will review the literature, including our own contribution, on human and animal models studies of the role of cytokines in learning and memory.

Chapter 1

INTRODUCTION

The interactions between the immune and the central nervous systems were described many years ago. One of the main discoveries was that peripheral interleukin (IL)-1 activates the hypothalamus-pituitary-adrenal (HPA) system [1]. The number of cytokines involved in brain-immune system interactions is now known to be huge [2, 3] and includes both pro-inflammatory and anti-inflammatory molecules. In recent years, the discovery of multiple functions of cytokines within the central nervous system (CNS) suggests that cytokines play a central role in the CNS function, independent of their role as messengers from the immune system to the brain.

Cytokines are a class of polypeptides that are expressed at low levels in healthy tissue but can be induced rapidly in response to tissue trauma or immune challenges to serve a variety of immune signaling and functions. More than 40 different cytokines have been identified to date, as well as numerous cytokine receptor types and subtypes. The knowledge of the functional role of cytokines has broadened since the characterization of the first cytokine, IL-1. Several authors have proposed the division of cytokines into "families" on the basis of functional or molecular characteristics.

Across classificatory schemes, members of each proposed cytokine family is either constitutively expressed or inducible in brain cells. The primary cellular sources of cytokines in the CNS are activated microglia and, to a lesser extent, astrocytes. Cytokines can also be synthesized in neurons, and small amounts may enter the brain from peripheral circulation. The brain-derived cytokines constitute the majority of the CNS cytokine pool and have predominantly paracrine and autocrine actions. They function as nerve growth factors crucial to embryogenesis and tissue repair and remodeling, as chemotactic factors, as mediators of phagocytosis, and as mediators of a variety of immunoregulatory functions in both the periphery and in the CNS.

The over-expression of cytokines has been associated with numerous pathological states (both within the CNS and in the peripheral nervous system), such as infection (viral, bacterial and fungal), autoimmune diseases (e.g., multiple sclerosis and systemic lupus erythematosus), stroke, trauma, neurodegenerative diseases (e.g., Alzheimer's, Parkinson's and prion diseases) and in neuropsychiatric disorders, such as depression and schizophrenia [4, 5]. More specifically, some cytokines such as interleukin-1β (IL-1β), interleukin-6 (IL-6) and tumor necrosis factor alpha (TNF-α) have been associated with cognitive decline and dementia in several cross-sectional and prospective population studies.

It is now believed that cytokine-mediated pathophysiological processes underlie the cognitive impairments associated with several neuropsychiatric diseases, making cytokines ideal targets for therapeutic interventions. Despite these pathological implications, cytokines have also been shown to exert physiological and even neuroprotective functions. The role of cytokines in modulating learning and memory, however, has been barely studied. Evidence from peripheral and central inflammation suggest that at least some pro-inflammatory cytokines can affect learning and memory formation. A few studies have evaluated the role of endogenous, normally expressed cytokines in these processes.

Chapter 2

SOURCES OF CYTOKINES IN THE BRAIN

Brain immunity is mediated predominantly by glial cells, particularly astrocytes (approximately 85% of the glial cell population) and microglia (approximately 10% of glial cells). Immune-activated glial cells produce a host of immune-signaling and effector molecules, including pro-inflammatory and anti-inflammatory cytokines, chemotactic and cell adhesion molecules, complement proteins, and cytotoxic free radicals. Neurons, too, have been demonstrated to produce some cytokines.

Astrocytes are cells from ectodermic origin that actively support normal physiological processes of neurons (e.g., in GABA and glutamate synthesis). Astrocytes can be activated (a process called "gliosis") in response to cytokines or other soluble molecules released during tissue injury, triggering astrocyte proliferation and increased synthesis and secretion of a range of immune effector molecules. Cytokines produced by astrocytes affect, in turn, the physiology of neighboring neurons and glial cells.

Microglia cells are endodermic cells that invade the developing brain and stay through life as resident macrophages in the CNS. Microglia is activated when surface receptors are engaged by mitogens, such as bacterial endotoxin or circulating immune factors including certain cytokines, prostanoids, and complement proteins. Activation can result in increased synthesis and secretion of soluble factors including cytokines, nitric oxide, reactive oxygen and nitrogen species, complement proteins, amino acid neurotransmitters and prostanoids.

Transmission of immune signals between the periphery and CNS may be a necessary event in the development of inflammatory brain damage in certain neurological disorders, including AIDS dementia, Multiple Sclerosis and

Alzheimer's disease. A pathogen entering the body elicits a rapid immune response mediated by macrophages. Recognition of lypopolysaccharide (LPS), a component of the cell wall of Gram-negative bacteria, by the Toll-like receptor 4 (TLR4), or the Gram-positive bacteria by TLR2, activates macrophages [6]. This activation results in the production of the pro-inflammatory cytokine IL-1, which is then able to induce its own synthesis and the synthesis of other cytokines that potentiate its effect (e.g., TNF-α and IL-6).

Immune cells and molecules in circulation can enter the brain directly through the blood-brain barrier [7, 8]. Cytokines are relatively large proteins, ranging in molecular weight from 8 to 40 kDa, and are hydrophilic, so they would not readily cross the blood-brain barrier. However, "activation" of brain endothelial cells by immune signals such as LPS or certain cytokines promotes interaction between circulating proteins and vascular endothelial cells by inducing expression of membrane adhesion molecules on endothelial cells [7-11]. Immune-activated brain endothelial cells can undergo changes in morphology, causing gaps to form in the wall of the blood-brain barrier [7, 10] and leakage of immune mediators into the brain.

Cytokines also enter the brain parenchyma at circumventricular organs—areas of the brain vasculature that lack tight cellular junctions—and modulation of neuroendocrine parameters at the organum vasculosum lamina terminalis (OVLT), a circumventricular organ adjacent to the hypothalamus, which has been attributed to the actions of circulating cytokines [12]. Finally, saturable, active transport of the cytokines IL-1α, IL-1β, IL-6 and TNF-α across the blood-brain barrier has been demonstrated in rodents [9, 13], although the active transport occurs at too low a level to account for the concentration of cytokines in the brain after peripheral administration of immune stimulants [2].

Although monocytes can produce pro-inflammatory cytokines that circulate in the blood and can reach far distant targets, most cytokines act locally in a paracrine manner. Neural afferents, such as the vagus nerves, can be the target of these cytokines. Sensory neurons of the vagus nerves express receptors for IL-1, and IL-1 stimulates vagal sensory activity [14]. Dantzer [15] and others [16, 17] have shown that vagus nerve afferents, stimulated by peripheral cytokines, can carry a neural code capable of triggering an inflammatory cascade in the brain. Sub-diaphragmatic lesion of the vagus nerve in rats attenuates brain cytokine production, corticosterone release and behavioral effects after i.v.-IL-1 administration [18, 19]. Vagotomy also blocks behavioral effects of intraperitoneal IL-1 [20, 21], and the induction of cytokines in the brain [22, 23]. Intraperitoneal IL-1, however, was efficient in inducing fever in vagotomized mice [24]. This evidence suggests that there are different mechanisms to communicate peripheral inflammation that result in different central responses.

Low levels of circulating cytokines may be instrumental in triggering cytokine production by immunocompetent glial cells in some cases of brain inflammation. However, numerous studies indicate that the neurotoxic consequences of brain inflammation are attributable to brain-derived, not peripheral, immune factors [3, 25, 26]. Cytokines and their receptors are expressed by brain cells, both glia and neurons [2, 27]. These endogenous molecules not only serve as communicators of immune signals into the brain and modulators of glial response, but they also appear to contribute to brain function.

Chapter 3

CYTOKINES IN DISEASE

SICKNESS BEHAVIOR

Animals and humans experiencing viral or bacterial infections usually exhibit various changes in behavior commonly referred as "sickness behavior" [28]. This behavior comprises malaise, weakness, listlessness, inability to concentrate, depression, lethargy, little interest in the surroundings, reduced appetite, numbness, reduced sexual activity and fatigue. These psychological and behavioral components of sickness represent, together with the fever response and the associated neuroendocrine changes, a highly organized strategy of the organism to fight infection [29]. The necessary synchrony between metabolic, physiological, and behavioral components of the systemic response to infection is dependent on the same molecular signals as those that are already responsible for the local inflammatory response. These signals are pro-inflammatory cytokines, such as interleukin-1 (IL-1), interleukin-6 (IL-6), tumor necrosis factor-α (TNF-α), and interferons (IFNs).

Cytokine-induced sickness behavior has had different interpretations since its first observation in the late 1980s [30]. In 1989, Dantzer and Kelley [31] proposed that the sickness-inducing properties of cytokines were not an artifact but reflected the genuine action of these proteins in the brain. They proposed that the immune system would require communication with the brain to regulate the host response to pathogens and that the brain would, in time, require communicating to the immune system. Hart [32] similarly saw sickness behavior as an adaptive response to reduce energy consumption at a time of increased energy demand that is necessary to maintain fever and to fight infection. However, it was later shown that IL-1-induced sickness behavior is independent of the febrile response [33]. A different view proposed that cytokines are responsible for the hyperalgesia that occurs during inflammation [34] and that the sickness response to cytokines is just an avatar of this necessity to deal with pain [35].

The current view of sickness behavior as the expression of a motivational state came only later [28]. This view took Hans Selye's proposal that infectious agents are stressors that trigger a counter-regulatory response aiming at re-establishing homeostasis. This way of defining cytokine-induced sickness behavior as a motivational state implies that it belongs in the realm of physiology, just as fear, hunger or thirst. As a consequence, sickness behavior needs to be studied independently of other phenomena such as fever.

Decreased social exploration of juvenile conspecifics has been used as a convenient way to assess sickness behavior in laboratory rodents. It offers the advantage of being reproducible and quantifiable. Systemic administration of LPS, IL-1β and TNF-α to adult rats or mice consistently decreases the time spent in exploration of juveniles [36-38].

Systemic administration of IL-1β and TNF-α also consistently suppresses feeding. This effect has been observed using various measurements of food intake under *ad libitum* as well as deprived conditions [39-41]. In contrast to the decrease in social exploration that takes about two hours to develop, the suppression of food intake occurs within one hour following treatment.

Exploiting a natural rodent behavior, the burrowing test has proved to be very useful in detecting sickness behavior even when fever is not observed [42]. Mice and rats spontaneously empty a tube filled with food pellets, gravel or other substances. This test is extremely sensitive to cytokines in rats and LPS in mice and rats.

The pathways of cytokine-induced sickness behavior in the brain are still under investigation. Several studies have demonstrated that peripheral cytokines induce the synthesis and release of cytokines in the brain. Intraperitoneal injection of LPS, for instance, induces the expression of IL-1α, IL-1β, and TNF-α, followed by that of IL-6 in many regions of the brain, including the hippocampus, the striatum and the thalamus [25, 43-45]. In this case, the main cellular sources of IL-1 are microglial cells and perivascular and meningeal macrophages [46].

Intraperitoneal challenge with the synthetic double-stranded RNA poly inosinic:poly cytidylic acid (poly I:C) results in alterations in core-body temperature, loss of body weight, reduced locomotor activity and reduced engagement in a species-typical behavior, namely burrowing [4]. These changes are accompanied by synthesis of IL-1β, IL-6, TNF-α and IFN-β in the hippocampus and the hypothalamus.

Administrating the IL-1 receptor antagonist (IL-1ra) into the lateral ventricle of the brain, it was shown that blocking brain IL-1 receptors abrogated the depressing effect of peripherally administered IL-1 on social exploration in rats [33]. This suggests that central cytokines induced by peripheral inflammation can act within the brain to elicit at least part of the behavioral effects associated with sickness.

Using expression of the early gene c-fos as a marker of neuronal activation, it was shown that intraperitoneal injection of LPS activates the primary projection area of the vagus nerve in the brain, the nucleus tractus solitarius, and the secondary projections of this nerve, including the parabrachial nucleus, the hypothalamic paraventricular and supraoptic nuclei, the central nucleus of the amygdala, and the bed nucleus of the stria terminalis [47]. A subdiaphragmatic vagotomy abrogated the expression of Fos in the brain after this intraperitoneal stimulus. The key role of the vagus nerve in the transmission of peripheral immune signals to the brain was further confirmed by the demonstration that vagotomy attenuates the behavioral actions of peripheral cytokines [18, 48].

Nadjar et al. [49] have shown that cerebral NFκB activation is a crucial step in the transmission of immune signals from the periphery to the brain. Injection of the NEMO Binding Domain peptide into the lateral ventricle blocked the behavioral effects of intraperitoneal IL-1β injection in the form of social withdrawal and decreased food intake. This treatment also abolished NFκB activation and Cox-2 synthesis in the brain microvasculature and reduced c-Fos expression in various regions of the brain. This report shows then that the intracellular NFκB signaling pathway is involved in IL-1β-induced sickness behavior.

The data presented so far suggests that peripheral inflammation elicits sickness behavior by activating the vagus nerve and its projections. This activation induces central expression of pro-inflammatory cytokines that, probably, lead to NFκB activation. Nevertheless, more research in trying to solve the mechanisms of peripheral inflammation inducing sickness behavior is needed.

As mentioned before, one feature of sickness behavior is depressed cognitive functioning. Using a behavioral task called autoshaping, Aubert et al. [50] demonstrated the existence of specific cognitive deficits in rats, independent of other effects of different pyrogens. Autoshaping consists of presenting hungry rats with a stimulus (introduction of a retractable lever) that predicted food delivery. Control rats quickly learned to press the lever, although this response does not influence the probability of food delivery. When LPS, IL-1 or yeast were injected i.p. to rats during acquisition of the task, treatment severely disrupted acquisition, as indexed by increased latencies of the conditioned response among treatment groups relative to controls. This deficit in acquisition persisted throughout 12 days of training, although the acute effects of immune stimulation persisted for fewer than 24 h. This suggests that treatment impaired learning in early trials, which affected performance throughout the remaining training period. In contrast to the effects of immune activation on autoshaping learning, stable autoshaped performance of a previously learned response was not affected by immune stimulation. This indicates a specific effect of immune activation on associative learning rather than on performance factors such as the anorectic or locomotor suppressive effects of inflammation. It is worth to notice, however, that these results might also be explained in terms of treatment-induced deficits in attention.

Peripheral administration of LPS impairs contextual fear conditioning [51]. Intraperitoneal injection of LPS in rats after the standard conditioning procedure resulted in impairment in hippocampal-dependent contextual fear conditioning. This treatment had no effect on hippocampal-independent auditory-cue fear conditioning. Moreover, using a variation of the fear conditioning paradigm, Pugh et al. [52] showed that LPS administered after context pre-exposure eliminated the pre-exposure beneficial effect on subsequent contextual fear conditioning. Pre-exposure is thought to facilitate contextual fear conditioning by allowing the rat to construct a memory representation of the context before conditioning occurs, a process thought to depend on the hippocampus [53].

Intraperitoneal injections of IL-1β daily for six days also has been shown to significantly impair spatial working memory in mice tested in the Morris water maze [54]. Mice injected with a sub-lethal dose of the gram-negative bacteria, *Legionella pneumophila* (Lp), also showed attenuated learning in the water maze. However, administration of Lp concurrently with intraperitoneal IL-1β antibodies (IL-1β Ab) that block the bioactive effects of IL-1β released by virus-stimulated immune cells, resulted in learning rates equal to saline-treated control mice. These results suggest that learning impairments produced by Lp infection are mediated in part by IL-1β released by stimulated immune cells. Although IL-1β Ab restored normal learning in Lp-infected mice, it did not block Lp-induced fever, again suggesting that the thermoregulatory and cognitive effects of inflammation are not linked. It is worth mentioning, however, that IL-1β-injected animals showed nonspecific symptoms of sickness, such as reduced activity and anorexia. As it was later shown for peripheral LPS, these treatments can have performance effects that make it difficult to specifically evaluate learning in tests like the Morris water maze, where swimming is required [55].

Banks et al. [56] showed that the CNS effect of IL-1α administered intravenously (i.v.) was dependent, in large part, on the ability of that cytokine to cross the BBB. They injected i.v. human IL-1α immediately after training and observed a significant impairment in retention on a T-maze footshock avoidance apparatus. It was previously shown that the transport of IL-1α across the blood-brain barrier is particularly high into the posterior division of the septum [57]. Injecting antibodies specific to human IL-1α directly into the posterior division of the septum, Banks et al. showed that memory impairment could be reversed. As these antibodies do not neutralize murine IL-1α, authors demonstrated that blood-borne IL-1α acts on the posterior division of the septum to block memory formation. Interestingly, i.v. human IL-1β also impaired retention of the task, but animals injected in the posterior division of the septum with neutralizing antibodies specific to human IL-1β showed the same level of impairment in retention than i.v. human IL-1β alone.

To examine the functional role of IL-6 in hippocampus-mediated cognitive impairments associated with peripheral inflammation, IL-6 KO mice were trained in a matching-to-place version of the water maze [58]. In wild-type mice, LPS administered after the training phase reduced the efficiency in mice locating the platform. IL-6 KO mice, on the contrary, appeared refractory to LPS-induced impairments in spatial working memory. Because authors observed a normal activation of the NTS after peripheral LPS, they concluded that humoral and neural immune-to-brain communication pathways appear to be independent of IL-6.

In summary, it has been shown that different peripheral inflammatory stimuli can impair acquisition and/or retention of hippocampal-dependent tasks. Although the precise mechanisms of this modulation have not been elucidated, these and similar studies have suggested that cytokines within the brain can modulate learning, independent of the other behavioral deficits observed during sickness behavior. Further studies injecting cytokines in the lateral ventricles or in specific regions of the brain have proven a specific role of certain cytokines in learning and memory formation. This evidence will be discussed below (Cytokines in health: Role in learning and memory).

NEURODEGENERATION

In addition to being induced after peripheral inflammation, cytokines have been detected in the brain during various neurodegenerative diseases, including AIDS Dementia Complex, brain trauma, Alzheimer's disease, Multiple Sclerosis, Human Prion Disease and Parkinson's Disease. In these pathologies, activated microglia and/or astrocytes have also been detected.

It is important to note that presence of activated glial cells does not imply that the disease is an immune disorder. Different stimuli can result in glial activation, and this in turn can elicit an immune response that could further damage the diseased brain. Some authors have proposed immune-mediated neurotoxicity as a unifying hypothesis of neurodegenerative diseases, in the sense that neurodegenerative syndromes of diverse etiology may share the feature of inflammatory neuronal damage as the final pathway leading to clinical impairments [59, 60]. We will discuss here the immune-mediated pathogenesis in Alzheimer's disease, because one of the main consequences of brain inflammation in this disease is to affect cognition.

Alzheimer's disease is a primary degenerative disease of the central nervous system, which is characterized by widespread cortical atrophy and enlargement of the ventricles [61]. The progression of Alzheimer's disease ultimately leads to dementia, and behavioral and cognitive impairments are often the basis for patients' initial medical referral leading to the diagnosis [62]. The histopathological hallmark of Alzheimer's disease is the presence of β-amyloid (Aβ) plaques forming insoluble aggregates in the cortex [61].

The Aβ plaques in the brains of Alzheimer's disease patients are invariably associated with activated microglia and microglial secretory products [63-67]. Aβ plaques consist of a central core of aggregated amyloid peptide surrounded by activated glial cells. Although Aβ plaques are the defining histopathological feature of Alzheimer's disease pathophysiology, the Aβ peptide has not consistently been demonstrated to be neurotoxic [64, 68] or to cause learning or memory deficits [69].

Moreover, the presence of activated microglia in the brains of Alzheimer's patients does not in itself imply that these cells contribute to the clinical neuropathology. Indeed, considering the phagocytic function of microglial cells, it is not surprising that these cells are present in the vicinity of amyloid plaques. Thus, the co-localization of activated microglia cells and senile plaques has been interpreted as a secondary consequence of Alzheimer's disease pathogenic processes.

In addition to the phagocytic response to extracellular Aβ, microglia may become activated to synthesize and secrete cytokines upon binding of Aβ to a second-messenger coupled membrane receptor [70]. In fact, the aggregating form of the Aβ peptide induces TNF-α production by a human microglia cell line, THP-1 [71]. Subfragments of the Aβ peptide have also been shown to induce or enhance microglial synthesis of other cytokines, including IL-1, as well as the complement protein C3 and nitric-oxide, and Aβ can induce increased intracellular Ca^{2+} concentrations in cultured rodent microglia [72].

Early onset Alzheimer's disease associated with genetic variants of the apolipoprotein gene may also involve Aβ and brain inflammation in the pathogenic cascade. The espilon3 variant of apolipoprotein E (ApoE3) has been shown to block Aβ-induced microglia activation *in vitro*, whereas ApoE4, coded by an ApoE allele, which is a known genetic risk factor for familiar early onset Alzheimer's disease [73], does not protect against microglial activation [74]. This finding suggests that the cognitive deficits and cortical neuropathology associated with the ApoE4 allele may be caused by inflammatory mechanisms.

Hsiao et al. [75] drew a more direct link among brain inflammation, amyloidosis and behavioral deficits using a transgenic mouse model of Alzheimer's-like amyloidosis based on over-expression of the amyloid precursor protein gene (APP mice). APP mice have multiple copies of the APP gene and over-express APP in the brain. By ten months of age, APP mice develop consolidated Aβ plaques analogous to Aβ plaques present in the brains of Alzheimer's patients. APP mice showed age-dependent impairments in spatial learning in the Morris water maze and deficits in spontaneous alternation in a T-maze. Moreover, the emergence of these behavioral deficits correlated with Aβ plaque formation. Amyloid plaques in the brains of APP mice were co-localized with activated astrocytes and glial cells, indicating the presence of brain inflammation. Although the effects of brain inflammation on amyloidosis and behavior were not explicitly examined in this study, the findings are consistent with the possibility that inflammatory processes contribute both to the development of Aβ plaques and to the cognitive/behavioral impairments observed in these mice.

During brain inflammation, cytokines are likely to contribute to the development of cognitive dysfunction, given their ability to modulate neurotransmitter and second-messenger signal transduction systems that are known to subserve certain cognitive processes. In this line, cytokines have been shown to affect noradrenalin, serotonin, acetylcholine (Ach), glutamate, nitric oxide (NO), and dopamine neurotransmission, and the cAMP, IP-3, and Ca2+-dependent intracellular transduction pathways [3, 76, 77].

There is some evidence that NO dysregulation is a factor in Alzheimer's disease and that NO dysregulation is related to CNS inflammation. Cerebrospinal fluid from Alzheimer's disease patients has significantly lower levels of nitrate, a metabolic product of NO, than cerebrospinal fluid from aged-matched controls [78]. Further, there is decreased staining of NAPDH-diaphorase, an index of NO synthase (NOS) levels, in the hippocampus of Alzheimer's disease patients, and the degree of NOS deficit has been found to be correlated with severity of cognitive deficits measured prior to death [79]. However, not all investigators have found significant alterations in NO activity in the brains of Alzheimer's disease patients [80, 81].

Disruption of brain NO concentrations has also been reported to produce Alzheimer's disease-like behavioral deficits in experimental animals, particularly on measures of learning and memory functions [82]. For example, intracranial administration of NOS blockers causes impairments in spatial working memory in the Morris water maze in mice and rats [83, 84] and impairments in learning in the radial arm maze in rats [85]. These behavioral effects of NO blockade are transient and can occur without permanent damage to neurons [82].

One challenge to this analysis is that where abnormalities in brain NO activity have been demonstrated in brain or cerebrospinal fluid of Alzheimer's patients, these have shown decreased NO levels, whereas stimulation of glial cells with cytokines *in vitro* increases NO release. Similarly, while disruptions in learning, memory and other behaviors have been shown in experimental animals after blockade of NO synthesis, there are no reports directly linking NO hyperfunction to behavioral impairment.

A partial explanation would be the different NO effects in young, healthy cells versus those already at risk. For example, Ohno et al. [86] showed deficits in working memory in rats after intrahippocampal injection of the NOS inhibitor, L-NAME, and, in a later paper [84], this same group showed that intrahippocampal injection of L-NAME prevented deficits in working memory when administered 24 h after induction of cerebral ischemia. Significantly, cerebral ischemia is known to activate glial cells and glial iNOS [87]. These opposite effects of NOS blockade on memory performance when administered after ischemic challenge or in healthy animals may reflect activation of different cellular sources of NO.

In summary, while the relationship among cytokine-mediated NO hyperfunction, the evidence of NO dysregulation in Alzheimer's disease brains, and the behavioral effects of NOS inhibition in experimental animals cannot be definitively resolved based on the data currently available, the findings reviewed above suggest that NO may be one pathogenic agent in immune-mediated cognitive dysfunction in Alzheimer's disease.

Numerous studies in experimental animals and humans have supported the theory that age-related deficits in acetylcholine (ACh) neurotransmission produce cognitive impairments consistent with Alzheimer's pathology [88]. This theory became known as the "cholinergic hypothesis" of Alzheimer's disease in the 1980s. The general finding underlying the cholinergic hypothesis is a decrease in Ach release from cholinergic neurons projecting to the basal forebrain, hippocampus, and, to a lesser extent, in other CNS cholinergic circuits [89]. Ach dysregulation has been associated with behavioral dysfunction in animal experiments using a variety of behavioral measures [90-92]. Intracerebroventricular injection of the carboxyl-terminal fragment of APP significantly impaired cued, spatial, and working memory performance of mice in Y-maze and water maze tasks and showed decreases in Ach levels in the cerebral cortex and hippocampus compared with the saline controls [93].

The decrease in cholinergic activity in the brain has been attributed to hyperactivity of a post-synaptic enzyme that terminates cholinergic synaptic transmission through hydrolysis of acetylcholine, acetylcholinesterase (AChE; [94]). AChE is overexpressed by neuritis associated with β-amyloid plaques in the Alzheimer's brain [95-97]. AChE, in turn, has been shown to regulate processing of the β-amyloid precursor protein [98] and to accelerate assembly of amyloid peptide into β-amyloid fibrils [99, 100], suggesting a link between AChE overexpression and β-amyloid formation. Overexpression of human AChE in neurons of transgenic mice produces progressive cognitive deterioration as assessed by the Morris water maze [101]. Therapeutic interventions based on increasing CNS Ach neurotransmission has been reported to ameliorate mental dysfunction in animal models of Alzheimer's pathology [89] and in human patients [102].

Several studies have linked the cytokine IL-2 to cholinergic neurotransmission in the rat brain [7, 103-106]. Under normal physiological conditions, brain concentration of IL-2 is low, and IL-2 is unlikely to play an important role in normal ongoing brain processes. However, during conditions of CNS inflammation, IL-2 and IL-2 receptors are strongly up-regulated and may reach concentrations sufficient to modify Ach neurotransmission with consequent effects on cognitive functioning and behavior [103]. In brain slices, IL-2 had no effect on spontaneous Ach release from neurons in hippocampus, cortex or striatum [103], however IL-2 did significantly suppress K^+-evoked Ach release in hippocampal and striatal slices. This effect of IL-2 was dose dependent and was detectable at relatively low concentrations in hippocampal slices. Hanisch et al. [104] also tested the ability of other cytokines to modulate Ach release from hippocampal, striatal, or cortical slices and found IL-1α, IL-1β, IL-3, IL-5, IL-6 and TNF-α all to be ineffective.

A more recent report, however, has shown a central role of IL-1 as the cytokine inducing a self-propagating role of neuronal injury and consequent microglia activation and further IL-1 overexpression [107]. In particular, authors showed that IL-1 increases AChE levels in the brain. Moreover, *in vitro* secreted APP fragments can activate microglia, which then secretes IL-1. This IL-1, in turn, promotes AChE expression, proving a pathway of inflammation-mediated alteration of neuronal function.

In summary, there are three types of evidence supporting the role of inflammatory processes in Alzheimer's disease pathology: (1) the co-regulatory and neurotoxic interactions among Aβ, proteases and protease inhibitors, and glia-derived inflammatory mediators; (2) the modulation of neurotransmitter and neuromodulator systems by inflammatory mediators that can be induced by amyloid proteins; and (3) the induction of Alzheimer's-like cognitive/behavioral pathology by immune stimulation. The theory of immunopathogenesis of Alzheimer's disease has clear therapeutic implications—specifically, that anti-inflammatory drugs taken at critical times in the disease process may protect against the development or progression of the disease [108, 109]. Indeed, several studies have already sought to assess the therapeutic potential of anti-inflammatory drugs for Alzheimer's disease.

For example, Rich et al. [110] found that Alzheimer's patients who had been taking non-steroidal anti-inflammatory drugs (NSAID) daily for 12 months had a shorter duration of illness relative to age and performed better on tests of cognitive functioning (including memory) than Alzheimer's disease patients without a history of consistent NSAID use. Further, patients taking NSAID showed a slower rate of cognitive decline over one year of periodic testing than did non-NSAID patients in this study.

Moreover, Rogers et al. [111] reported similar findings in a six-month, double-blind prospective study comparing Alzheimer's patients treated with the potent anti-inflammatory drug, indomethacin, with matched, placebo-treated control subjects. By the end of six months, indomethacin-treated patients showed slight overall improvements in measures of cognitive functioning, whereas control subjects showed significant cognitive decline.

Chapter 4

CYTOKINES IN HEALTH: ROLE IN LEARNING AND MEMORY

INTERLEUKIN-1 (IL-1)

The interleukin-1 (IL-1) is a cytokine produced by a number of peripheral cell types and plays a variety of roles in immune and inflammatory responses [112-114]. IL-1 action is regulated by a complex network of molecules that includes multiple ligands (IL-1α, IL-1β, and the endogenous IL-1 receptor antagonist [IL-1ra]), several binding sites (IL-1 receptor [IL-1R] type I, IL-1R type II, IL-1 accessory protein, soluble receptors), and a key regulatory enzyme (IL-1-converting enzyme). As mentioned before, most of these molecules have been localized in the brain under varying conditions, both in neurons and in glial cells [115, 116].

The first unequivocal evidence showing that IL-1 was active in the brain was published in 1986 [1]. It was shown that administration of subpyrogenic doses of both purified and recombinant IL-1 to mice and rats increased blood levels of adrenocorticotropic hormone (ACTH) and glucocorticoids. Because the same effect was found in athymic nude mice, authors concluded that the effect observed supported the existence of an immunoregulatory feedback circuit on the hypothalamic-pituitary-adrenal (HPA) axis. In fact, they later showed that IL-1 activates the corticotrophin-releasing factor containing neurons in the paraventricular nucleus of the hypothalamus [117].

It was subsequently shown that exogenously applied IL-1 can modulate neuronal activity inhibiting calcium influx [118, 119], protein kinase C [118], or the release of acetylcholine [120] and glutamate [118] in the normal hippocampus. In addition, IL-1 modulates a process related to learning, such as long-term potentiation (LTP) *in vivo* [121-126]. In this model, as in other models in the periphery, physiological low levels of IL-1 could have opposite effects from those observed when larger amounts are induced or applied exogenously [51].

Oitz et al. [127] reported for the first time impaired memory performance in rats in the Morris water maze after treatment with proinflammatory cytokines. IL-1β or IL-6 were injected into the cerebral ventricles of rats 60 minutes prior to water maze training, using a procedure designed to test spatial learning. While treatment with IL-6 had no effect, water maze performance of IL-1β-treated rats was significantly impaired relative to saline treated controls. Both IL-6 and IL-1β significantly increased body temperature, suggesting that cognitive effects of IL-1β are not a consequence of the immune thermoregulatory response.

In the same line, Pugh et al. [128] showed that administration of the HIV protein gp120 into the lateral ventricle immediately following fear conditioning, impaired contextual fear conditioning, while having no effect on auditory-cue fear conditioning. As gp120 induces IL-1β mRNA in the brain [129], authors tested whether IL-1 antagonism could block this effect on memory. IL-1ra i.c.v. given after conditioning but before gp120 administration blocked the effect of gp120 on conditioning.

Matsumoto et al. [130] studied the effect of exogenous IL-1β on working memory performance of rats. Using a three-panel runaway task, the authors showed that IL-1β injected laterally into the dorsal hippocampus significantly increased the working memory errors in this test, while the number of errors in the first trial or the latencies to obtain food pellets were not affected. Moreover, the combined injection of IL-1β and IL-1ra into the hippocampus significantly decreased the working memory deficit. Finally, authors proposed that IL-1β effect is mediated by acetylcholine and/or glutamate in the hippocampus, as both a cholinesterase inhibitor and a NMDA partial agonist revert IL-1β-induced working memory deficits.

Pre-exposure of rats to the context has a facilitatory effect on contextual fear conditioning 24 h later. Barrientos et al. [131] showed that IL-1β injected into the dorsal hippocampus after pre-exposure to a conditioning context, interferes with the context pre-exposure facilitation effect. Interestingly, IL-1β blocks context conditioning independently of corticosteroid induction in the periphery. These results suggest that intra-hippocampal IL-1β interferes with memory consolidation of contextual representation. Moreover, they showed that IL-1β can affect memory consolidation both immediately after context exposure and three or 24 h later.

Using a different approach, Palin et al. [132] studied the role of IL-1 in the memory impairment associated with a second exposure to bacillus Calmette-Guerin (BCG). Animals were primed with an injection of BCG into the dorsal hippocampus and challenged four weeks later with BCG administered subcutaneously. This challenge was associated with impaired spatial memory in a Y-maze and with increases of IL-1β levels in the hippocampus. The chronic subcutaneous administration of IL-1ra abrogated these effects. These results show that increases of IL-1β during inflammation can result in cognitive impairments.

Trying to identify the molecular mechanisms of IL-1 modulation of learning, Hein et al. [133] showed that prostaglandins are necessary and sufficient to impair contextual fear conditioning after IL-1β treatment. Cox-2 was induced after IL-1β injection into the dorsal hippocampus, and prostaglandin E_2 (PGE_2) reduced freezing during testing. Finally, PGE_2 injected during the two hours post-conditioning reduced the conditioning-induced increase in brain-derived neurotrophic factor (BDNF), a molecule previously shown to be necessary after training to form a long-term memory. Therefore, IL-1 can regulate memory formation through the induction of neurotrophins.

Although these reports demonstrated that IL-1 could play a role in learning and memory processes during inflammation in the brain, they did not address the question of whether endogenous hippocampal IL-1 modulates memory acquisition and consolidation during nonpathological situations.

To examine the hypothesis that IL-1 signaling is involved in learning and memory, Yirmiya et al. [134] injected rats i.c.v. with IL-1ra and tested their performance in the Morris water maze (MWM) and passive avoidance test. They found that IL-1ra caused memory impairment in the passive avoidance test and the standard version of the MWM, which assesses spatial memory. It is interesting to note that even though these authors analyze hippocampus-dependent behavior, they inject the IL-1ra into the ventricles. I.c.v. injection distributes proteins to different regions of the brain and has strong effects on the ventricles walls [135]. In fact, among generalized effects on behavior, authors report an increase in anorexia. However, the influence of these unspecific effects on learning has not been analyzed.

Avital et al. [136] showed that previously generated type I IL-1 receptor knockout mice (IL-1rKO; [137]) show learning deficiencies. IL-1rKO mice displayed a slower rate of learning in the spatial memory version of the MWM and impaired contextual fear conditioning. In contrast, IL-1rKO mice showed normal hippocampal-independent memory in the visually guided task of the water maze and in the auditory-cued fear conditioning. However, IL-1rKO mice show slower speed of swimming, suggesting that they have motor impairments and reduced pain sensibility [138]. This suggests that IL-1rKO mice not only have compromised hippocampal plasticity but that also other systems are affected.

To study the role of IL-1 system in learning, we chose a fear-motivated learning task that is hippocampal-dependent and is acquired in a single and brief training session [139]. We first analyzed the expression of IL-1α, IL-1β, and IL-1ra during the posttraining time window previously shown to be involved in the memory consolidation process [140-142]. We observed a transient increment in IL-1α mRNA levels four hours after training. The IL-1α increment observed was specific for the training and was not related to the stress of the procedure, because control animals submitted to the same shock and handling, but incapable of making the stepping down-shock association, showed no alterations of their IL-1α mRNA levels. In our experiments, IL-1β mRNA levels were not altered.

As described before, IL-1 expression is induced by stress in some brain areas but detected only in adrenalectomized animals, since glucocorticoids downregulates IL-1α levels in shocked animals [143]. It is interesting to highlight that, in our experiments, IL-1α expression was refractory to the inhibitory action of stress through glucocorticoids, strengthening a possible specific role of the *de novo* production of this cytokine in learning. Thus, these results showed a learning-specific regulation of IL-1α, suggesting that IL-1α could be involved in memory formation. This prompted us to further investigate the role of endogenous IL-1α in hippocampal memory processing.

Using an adenoviral vector that expresses the natural antagonist of the IL-1 receptor (IL-1ra), we were able to block IL-1 function specifically in the dorsal hippocampus and to show that this blockade has a facilitatory effect on short-term memory (STM) and long-term memory (LTM) in an inhibitory avoidance learning task. These results cannot be explained by the IL-1α capability of modulating nociceptive information [144, 145], because all experimental groups had similar current thresholds for jumping and vocalization [146, 147]. Moreover, we showed that IL-1ra expression in the dorsal hippocampus had no effect on the number of crossings and rearings in an open field, so that the adenoviral vector does not affect locomotor activity. Given that the increase in stepping-down latency could not be due to increased shock sensitivity or to the reduction of locomotor activity, our results strongly suggest that blockade of hippocampal IL-1 receptor has a learning-specific effect on the inhibitory avoidance learning task.

However, given that animals were trained and tested under the influence of AdIL-1ra, we were unable to elucidate whether endogenous IL-1 is involved in acquisition, consolidation, and/or retrieval during hippocampal memory processing. Some reports have emphasized the physiological importance of learning and memory suppressors [148]. Thus, we postulated that endogenous IL-1α in the hippocampus could act as a modulator to constrain physiologically memory formation or retrieval, or both. However, this does not rule out that different forms of memory can be modulated differently by IL-1 or by different IL-1 levels, so that endogenous IL-1 could actually be facilitatory in other learning tasks.

To evaluate whether IL-1 receptor blockade had similar effects on a nonassociative learning task [149-151], we evaluated AdIL-1ra animals during one of the most elementary nonassociative learning tasks, i.e., the behavioral habituation to a novel environment. In this paradigm, the blockade of IL-1 receptor had no effect on LTM. So we demonstrated that two hippocampal-dependent memories, one requiring association to an aversive stimulus and other nonassociative, have different sensitivity to IL-1. Namely, IL-1ra facilitates the inhibitory avoidance task but has no effect in the habituation to a novel environment.

Taken together, our results suggest a specific role of the endogenous proinflammatory cytokine IL-1α in the formation and/or retrieval of an associative memory. This shows a modulatory function of a cytokine similar to those previously shown for neurotrophins (e.g., [140]), neurotransmitters and hormones [152, 153].

Goshen et al. [154] also studied the physiological role of IL-1 in learning and memory. As we showed before, authors found that learning of a hippocampal-dependent task induces IL-1 gene expression in the hippocampus. Mice that express the IL-1ra under the GFAP promoter (Tg IL-1ra mice) show disrupted hippocampal-dependent memory in both contextual-fear conditioning paradigm and MWM. It was later shown that Tg IL-1ra mice present changes in BDNF-ERK1/2 and Src signalling pathways, suggesting an underlying mechanisms of IL-1 modulation of memory formation [155]. Considering their results, Goshen et al. suggested an inverted U-shaped pattern of IL-1 modulation of memory processes, i.e., physiological levels of IL-1, are needed for memory formation, and a slight increase in brain IL-1 levels can even improve memory, whereas any deviation from physiological range, either by excess elevation in IL-1 or by blockade of IL-1 signaling, results in impaired memory.

Genetic studies in humans showed that individuals carrying interleukin-converting enzyme polymorphisms that are accompanied by lower IL-1β levels (e.g., 10643C and 5352A) performed better on all cognitive tests [156]. A biallelic functional polymorphism in the promoter region (position -511) of the IL-1β gene is associated with cognitive performance in elderly males [157]. C/C genotype showed better cognitive function test results than T/T carriers. This latter group secretes significantly more IL-1β than -511C carriers [158], suggesting again that high IL-1 levels are detrimental to cognitive performance.

In summary, there is consensus that high IL-1 expression in the brain (and particularly in the hippocampus) blocks learning and memory consolidation of hippocampus-dependent tasks. The research has focused mainly on IL-1β, while the specific role of pathological levels of IL-1α has not been evaluated.

The role of endogenous, physiological levels of IL-1 is, however, less clear. The analysis of IL-1R KO mice and GFAP TgIL-1ra mice suggest that IL-1 may be necessary for learning and/or memory formation. However, these mice lack IL-1 signalling from conception or early development, and other unspecific effects cannot be ruled out. In fact, motor and nociceptive deficits have been reported. To our knowledge, our paper is the only one reporting a facilitatory effect of endogenous IL-1 blockade. Two particular facts are worth mentioning. First, the treatment with adenoviral vectors expressing IL-1ra is limited to the adult animal and the dorsal hippocampus. This is a clear difference with knockout and transgenic mice. Second, in our paradigm IL-1α, and not IL-1β, appeared involved in memory consolidation. These differences could account for the otherwise controversial results. Moreover, species-specific differences could also be cited.

INTERLEUKIN-6 (IL-6)

IL-6 regulates immune responses, acute phase reactions and hematopoiesis, and plays a central role in host defense mechanisms. IL-6 is a pleiotropic cytokine that is produced in a variety of cells and acts on a wide range of tissues, exerting growth-inducing, growth-inhibitory and differentiation-inducing effects, depending on the nature of the target cells [159]

IL-6 is expressed in the embryonic cortex and adult hippocampus of the rat brain [160]. Also IL-6 receptor (IL-6R) mRNA has been detected in the adult hippocampus and striatum [161, 162], specifically in the granular layer of the dentate gyrus, in the fimbria, and in large pyramidal neurons of the CA1 to CA4 hippocampal fields. Moreover, both IL-6 and IL-6R mRNAs are expressed in the granular cell layer of the cerebellum. Membrane depolarization induces IL-6 mRNA accumulation in primary cortical cells and is increased *in vivo* after epileptic seizures [163]. However, IL-6 can also enhance the survival of several classes of neurons in culture [164, 165].

IL-6 exerts pronounced effects on the CNS: it stimulates neuronal growth [166, 167], reduces food intake [168], induces fever [169] and activates the hypothalamic-pituitary-adrenal (HPA) system [170]. Transgenic mice with high levels of IL-6 expression develop severe neurologic disease characterized by tremor, ataxia, and seizure [171]. These mice also exhibit neurodegeneration, astrocytosis, and angiogenesis.

Heyser et al. [172] used a transgenic mouse model of chronic CNS inflammation to study the effects of brain inflammation on spatial learning in a Y-maze. Transgenic mice (GFAP-IL-6 mice) express IL-6 chronically in astrocytes, due to fusing the gene coding for IL-6 to the promoter of the normally expressed astrocytic protein, glial fibrillary acidic protein GFAP. The brains of GFAP-IL-6 mice are characterized by localized inflammation and localized, age-dependent neurodegeneration in the hippocampus and cerebellum, as well as increased blood-brain barrier permeability. GFAP-IL-6 mice also show microglia activation and increased cytokine production by microglia and astrocytes in the hippocampus. While GFAP-IL-6 mice aged less than six months showed normal avoidance learning, GFAP-IL-6 mice aged six months or older were found to be significantly impaired in a Y-maze active avoidance learning task compared to wild-type control mice. The age dependence on learning deficits in GFAP-IL-6 mice suggests a temporal accumulation of inflammatory neurotoxicity, and, according to the authors, represents a plausible model for the progressive degenerative behavioral disorders in Alzheimer's and other human dementias involving chronic CNS inflammation.

In line with these results, Braida et al. [173] found facilitation in learning over 30 days in IL-6 KO mice. IL-6 KO mice showed reduced number of working memory errors in a radial maze. Moreover, these mice showed normal spontaneous motor activity, suggesting that the effect is not due to unspecific differences in motor function. This revealed a facilitatory effect of

IL-6 gene deletion on spatial learning, as opposed to the detrimental effect of IL-6 over-expression demonstrated before.

However, other authors have found impaired memory in IL-6 KO mice, both in the MWM and in the novel object recognition test [174]. In the MWM, IL-6 KO mice displayed both reduced acquisition and impaired reversal learning. Moreover, IL-6 KO mice showed reduced exploratory behavior.

The disturbance of IL-6 signaling in IL-6 KO mice is very likely to affect brain development and the development of the immune system and its connections with the brain. As a consequence, neuronal functions correlated with memory processes would be compromised. Furthermore, disturbed IL-6 signaling leads to behavioral changes (reduced exploratory behavior, increased stress susceptibility) that have negative impact on memory formation. In conclusion, new tools should be generated that block IL-6 signaling specifically in the brain and in a restricted period of time.

Human studies have shown that high IL-6 plasma levels are associated with risk in decline of cognitive function in elderly people [175]. Moreover, a study among newborns reported a relationship between specific IL-6 genotypes and cognitive function showing that the 572C-allele of IL-6 (CC/GC genotypes) is associated with impaired cognitive development among children [176].

In summary, although there is less work studying the role of IL-6 in learning and memory in comparison with IL-1, all evidence suggests that increased IL-6 into the brain is detrimental to learning. Again, the role of endogenous IL-6 is less clear, some evidence showing that blockade of basal IL-6 signaling facilitates learning and other suggesting that basal IL-6 activity is required to learn.

TUMOR NECROSIS FACTOR-ALPHA (TNF-α)

TNF-α is a cytokine produced by activated cells of the immune system, mainly by macrophages [177]. It is produced mostly as a 17 kD secreted protein, although a membrane-bound variant of the protein has citotoxic activity and an important role in cellular communication. There are two types of TNF receptors, TNF-RI (55-60 kDa) and TNF-RII (75-80 kDa), which can be shed, therefore existing in soluble form [178]. TNF-α binds both receptors with approximately equal affinity. These two distinct receptors have been shown to elicit different molecular and cellular responses [179-181]. For example, binding to the TNF-RI initiates the activation of caspases and can ultimately lead to apoptosis [181]. Conversely, TNF-RII appears to inhibit the activation of caspases and protect against apoptosis. TNF-α production in the periphery is induced by different stimuli, including mitogens, IL-1, IL-2 and interferons.

TNF-α is expressed at very low levels in basal conditions in the central nervous system. In contrast, astrocytes, microglia and neurons have been described to produce TNF-α upon exposure to multiple physiological and pathological stimuli (e.g., CNS infection and brain injury) [178]. In the brain, as opposed to what has been described in the periphery, IL-1 fails to induce TNF-α expression, and TNF-α does not induce either itself or IL-1 [182, 183].

Transgenic mice, specifically overexpressing murine TNF-α in the CNS either in neurons or in astrocytes, and at different stages of development, spontaneously develop chronic CNS inflammation. This is characterized by widespread reactive astrocytosis and microgliosis, meningeal and parenchymal infiltration by activated T-cells and macrophages, loss of white matter and neurodegeneration [184]. On the contrary, mice lacking TNF-α or both receptors show exacerbated neuronal damage elicited by different neurotoxic stimuli [185, 186]. Therefore, TNF-α can be neurotoxic or have neuroprotective effects depending on level of expression, type of neurons affected and other factors.

TNF-α blocks LTP in slices [121], at least partly through p38 MAPK and metabotropic glutamate receptors [187, 188]. It has been suggested that the continual presence of TNF-α is required for preservation of synaptic strengths at the excitatory synapse [189] and that neurotransmitter metabolism and neurodevelopmental processes were altered by TNF-α [190].

As mentioned before, TNF-α transgenic mice that express the cytokine in the brain develop neurologic symptoms, including CNS inflammation and demyelination [184]. These mice show a retardation of passive avoidance acquisition, with no effects during the retention phase [191]. However, as these mice show reduced body weight, reduced exploratory activity and analgesia, the specificity of the learning deficits is not clear.

Comparing that same line with a different transgenic line that shows no neurological or phenotypical alterations, Aloe et al. [192] showed that both lines exhibit impairments in learning in the MWM. More specifically, both transgenic lines showed increased escape latency during the test. It is worth mentioning that both lines exhibited reduced swimming speed, while the distance traveled was the same. This suggests that transgenic mice were less effective than wild type animals in using the acquired positional information either because of forgetting or because of motivational changes caused by the new stressing environmental situation. Alternatively, transgenic mice could be more distractible in learning a task, which requires more complex informational processing as retrieving the platform position. These mice also present reduced neurotrophic factors levels in the hippocampus (namely, NGF and BDNF), suggesting a possible mechanism of this effect. Finally, learning deficits in TNF-α Tg mice appear only at 30 to 35 days of age [193], suggesting that this phenotype might be influenced by the TNF-α produced in specific brain regions during the adult life.

Again, the role of the endogenous cytokine levels is less clear. One report showed that TNF-α KO mice have no phenotype in the Y-maze or the MWM [194], suggesting that this cytokine is not necessary for memory formation in these tests. In contrast, other work demonstrated that TNF-α KO mice show a deficiency in spatial memory when young, but with age, TNF deficiency leads to facilitation in spatial memory [195].

A significant association between the TNF-α 308G to A polymorphism and processing speed was observed, indicating better performance for heterozygous or homozygous carriers of the A allele in elderly people [196]. This polymorphism, however, was not associated with altered cytokine plasma levels.

Chapter 5

Conclusion

Pro-inflammatory cytokines are expressed within the brain in health and in disease. All evidence suggests that IL-1, IL-6 and TNF-α can be induced in the brain under various pathological conditions and that this up-regulation results in detrimental effects on learning and memory. This is normally observed during sickness, when cognitive impairment is associated with a need to save energy to fight infection. However, neuropathological conditions that lead to increased cytokine production in the brain can also result in cognitive impairment. For example, in Alzheimer's disease, amyloid plaques are accompanied by gliosis, cytokine increments and cognitive impairment.

The role of low, endogenous levels of brain cytokines in learning and memory is less clear. Only few reports have studied the role of pro-inflammatory cytokines under these conditions, and results are controversial. It is essential to develop new genetic tools to elucidate the role of cytokines in the normal brain, without the confounding effects of inflammation.

ABBREVIATIONS

ACTH	adrenocorticotropic hormone
APP	amyloid precursor protein
CNS	central nervous system
GFAP	glial fibrillary acidic protein
HPA	hypothalamus-pituitary-adrenal
i.c.v.	intracerebroventricular
i.p.	intraperitoneal
i.v.	intravenous
IL-1	interleukin 1
IL-1R	IL-1 receptor
IL-6	interleukin 6
IL-6R	IL-6 receptor
KO	knockout
LPS	lipopolysaccharide
LTM	long-term memory
LTP	long-term potentiation
MWM	Morris water maze
NO	nitric oxide
STM	short-term memory
TNF-alpha	tumor necrosis factor alpha
TNF-R	TNF-alpha receptor

REFERENCES

[1] Besedovsky, H; del Rey, A; Sorkin, E; Dinarello, CA. Immunoregulatory feedback between interleukin-1 and glucocorticoid hormones. *Science*, 1986. 233(4764):652-654.

[2] Hopkins, SJ; Rothwell, NJ. Cytokines and the nervous system I: Expression and recognition. *Trends in Neurosciences*, 1995. 18(2):83-88.

[3] Rothwell, NJ; Hopkins, SJ. Cytokines and the nervous system II: Actions and mechanisms of action. *Trends in Neurosciences*, 1995. 18(3):130-136.

[4] Cunningham, C; Campion, S; Teeling, J; Felton, L; Perry, VH. The sickness behaviour and CNS inflammatory mediator profile induced by systemic challenge of mice with synthetic double-stranded RNA (poly I:C). *Brain, Behavior, and Immunity*, 2007. 21(4):490-502.

[5] Gilmore, JH; Jarskog, LF. Exposure to infection and brain development: cytokines in the pathogenesis of schizophrenia. *Schizophr Res*, 1997. 24(3):365-367.

[6] Akashi, S; Shimazu, R; Ogata, H; Nagai, Y; Takeda, K; Kimoto, M; Miyake, K. Cutting edge: cell surface expression and lipopolysaccharide signaling via the toll-like receptor 4-MD-2 complex on mouse peritoneal macrophages. *The Journal of Immunology*, 2000. 164(7):3471-3475.

[7] Carlson, SL; Beiting, DJ; Kiani, CA; Abell, KM; McGillis, JP. Catecholamines decrease lymphocyte adhesion to cytokine-activated endothelial cells. *Brain, Behavior, and Immunity*, 1996. 10(1):55-67.

[8] Merrill, JE; Murphy, SP. Inflammatory events at the blood brain barrier: regulation of adhesion molecules, cytokines, and chemokines by reactive

nitrogen and oxygen species. *Brain, Behavior, and Immunity*, 1997. 11(4):245-263.

[9] Banks, WA; Plotkin, SR; Kastin, AJ. Permeability of the blood-brain barrier to soluble cytokine receptors. *Neuroimmunomodulation*, 1995. 2(3):161-165.

[10] Ellison, MD; Povlishock, JT; Merchant, RE. Blood-brain barrier dysfunction in cats following recombinant interleukin-2 infusion. *Cancer Res*, 1987. 47(21):5765-5770.

[11] Saija, A; Princi, P; Lanza, M; Scalese, M; Aramnejad, E; De Sarro, A. Systemic cytokine administration can affect blood-brain barrier permeability in the rat. *Life Sci*, 1995. 56(10):775-784.

[12] Luheshi, GN; Hammond, E; Van Dam, AM. Cytokines as messengers of neuroimmune interactions. *Trends in Neurosciences*, 1996. 19(2):46-47.

[13] Banks, WA; Farr, SA; Morley, JE. Entry of blood-borne cytokines into the central nervous system: effects on cognitive processes. *Neuroimmunomodulation*, 2002. 10(6):319-327.

[14] Ek, M; Kurosawa, M; Lundeberg, T; Ericsson, A. Activation of vagal afferents after intravenous injection of interleukin-1beta: role of endogenous prostaglandins. *The Journal of Neuroscience*, 1998. 18(22):9471-9479.

[15] Dantzer, R; Bluthe, RM; Laye, S; Bret-Dibat, JL; Parnet, P; Kelley, KW. Cytokines and sickness behavior. *Ann N Y Acad Sci*, 1998. 840:586-590.

[16] Maier, SF; Goehler, LE; Fleshner, M; Watkins, LR. The role of the vagus nerve in cytokine-to-brain communication. *Ann N Y Acad Sci*, 1998. 840:289-300.

[17] Toulmond, S; Parnet, P; Linthorst, ACE. When cytokines get on your nerves: cytokine networks and CNS pathologies. *Trends in Neurosciences*, 1996. 19(10):409-410.

[18] Bluthe, RM; Michaud, B; Kelley, KW; Dantzer, R. Vagotomy attenuates behavioural effects of interleukin-1 injected peripherally but not centrally. *NeuroReport*, 1996. 7(9):1485-1488.

[19] Fleshner, M; Goehler, LE; Schwartz, BA; McGorry, M; Martin, D; Maier, SF; Watkins, LR. Thermogenic and corticosterone responses to intravenous cytokines (IL-1beta and TNF-alpha) are attenuated by subdiaphragmatic vagotomy. *Journal of Neuroimmunology*, 1998. 86(2):134-141.

[20] Bluthe, RM; Michaud, B; Kelley, KW; Dantzer, R. Vagotomy blocks behavioural effects of interleukin-1 injected via the intraperitoneal route

but not via other systemic routes. *NeuroReport*, 1996. 7(15-17):2823-2827.
[21] Bret-Dibat, JL; Bluthe, RM; Kent, S; Kelley, KW; Dantzer, R. Lipopolysaccharide and interleukin-1 depress food-motivated behavior in mice by a vagal-mediated mechanism. *Brain, Behavior, and Immunity*, 1995. 9(3):242-246.
[22] Hansen, MK; Taishi, P; Chen, Z; Krueger, JM. Vagotomy blocks the induction of interleukin-1beta (IL-1beta) mRNA in the brain of rats in response to systemic IL-1beta. *The Journal of Neuroscience*, 1998. 18(6):2247-2253.
[23] Laye, S; Bluthe, RM; Kent, S; Combe, C; Medina, C; Parnet, P; Kelley, K; Dantzer, R. Subdiaphragmatic vagotomy blocks induction of IL-1 beta mRNA in mice brain in response to peripheral LPS. *American Journal of Physiology*, 1995. 268(5 Pt 2):R1327-1331.
[24] Luheshi, GN; Bluthe, RM; Rushforth, D; Mulcahy, N; Konsman, JP; Goldbach, M; Dantzer, R. Vagotomy attenuates the behavioural but not the pyrogenic effects of interleukin-1 in rats. *Auton Neurosci*, 2000. 85(1-3):127-132.
[25] Ban, E; Haour, F; Lenstra, R. Brain interleukin 1 gene expression induced by peripheral lipopolysaccharide administration. *Cytokine*, 1992. 4(1):48-54.
[26] Perry, V; Andersson, P-B; Gordon, S. Macrophages and inflammation in the central nervous system. *Trends in Neuroscience*, 1993. 16(7):268-273.
[27] Rothwell, NJ; Luheshi, G; Toulmond, S. Cytokines and their receptors in the central nervous system: physiology, pharmacology, and pathology. *Pharmacol Ther*, 1996. 69(2):85-95.
[28] Dantzer, R. Cytokine-induced sickness behavior: mechanisms and implications. *Ann N Y Acad Sci*, 2001. 933:222-234.
[29] Konsman, JP; Parnet, P; Dantzer, R. Cytokine-induced sickness behaviour: mechanisms and implications. *Trends in Neurosciences*, 2002. 25(3):154-159.
[30] Dantzer, R; Kelley, KW. Twenty years of research on cytokine-induced sickness behavior. *Brain, Behavior, and Immunity*, 2007. 21(2):153-160.
[31] Dantzer, R; Kelley, KW. Stress and immunity: an integrated view of relationships between the brain and the immune system. *Life Sci*, 1989. 44(26):1995-2008.
[32] Hart, BL. Biological basis of the behavior of sick animals. *Neurosci Biobehav Rev*, 1988. 12(2):123-137.

[33] Kent, S; Bluthe, RM; Dantzer, R; Hardwick, AJ; Kelley, KW; Rothwell, NJ; Vannice, JL. Different receptor mechanisms mediate the pyrogenic and behavioral effects of interleukin 1. *Proc Natl Acad Sci USA*, 1992. 89(19):9117-9120.

[34] Maier, SF; Wiertelak, EP; Martin, D; Watkins, LR. Interleukin-1 mediates the behavioral hyperalgesia produced by lithium chloride and endotoxin. *Brain Research*, 1993. 623(2):321-324.

[35] Watkins, LR; Maier, SF. The pain of being sick: implications of immune-to-brain communication for understanding pain. *Annu Rev Psychol*, 2000. 51:29-57.

[36] Bluthe, RM; Crestani, F; Kelley, KW; Dantzer, R. Mechanisms of the behavioral effects of interleukin 1. Role of prostaglandins and CRF. *Ann N Y Acad Sci*, 1992. 650:268-275.

[37] Bluthe, RM; Dantzer, R; Kelley, KW. Effects of interleukin-1 receptor antagonist on the behavioral effects of lipopolysaccharide in rat. *Brain Research*, 1992. 573(2):318-320.

[38] Bluthe, RM; Pawlowski, M; Suarez, S; Parnet, P; Pittman, Q; Kelley, KW; Dantzer, R. Synergy between tumor necrosis factor alpha and interleukin-1 in the induction of sickness behavior in mice. *Psychoneuroendocrinology*, 1994. 19(2):197-207.

[39] Kent, S; Bret-Dibat, JL; Kelley, KW; Dantzer, R. Mechanisms of sickness-induced decreases in food-motivated behavior. *Neurosci Biobehav Rev*, 1996. 20(1):171-175.

[40] Plata-Salaman, CR. Cytokines and feeding suppression: an integrative view from neurologic to molecular levels. *Nutrition*, 1995. 11(5 Suppl):674-677.

[41] Plata-Salaman, CR. Cytokines and feeding. *Int J Obes Relat Metab Disord*, 2001. 25 Suppl 5:S48-52.

[42] Deacon, RM. Burrowing: a sensitive behavioural assay, tested in five species of laboratory rodents. *Behavioural Brain Research*, 2009. 200(1):128-133.

[43] Gatti, S; Bartfai, T. Induction of tumor necrosis factor-alpha mRNA in the brain after peripheral endotoxin treatment: comparison with interleukin-1 family and interleukin-6. *Brain Research*, 1993. 624(1-2):291-294.

[44] Laye, S; Parnet, P; Goujon, E; Dantzer, R. Peripheral administration of lipopolysaccharide induces the expression of cytokine transcripts in the brain and pituitary of mice. *Brain Res Mol Brain Res*, 1994. 27(1):157-162.

[45] Pitossi, F; del Rey, A; Kabiersch, A; Besedovsky, H. Induction of cytokine transcripts in the central nervous system and pituitary following peripheral administration of endotoxin to mice. *Journal of Neuroscience Research*, 1997. 48(4):287-298.

[46] van Dam, AM; Brouns, M; Louisse, S; Berkenbosch, F. Appearance of interleukin-1 in macrophages and in ramified microglia in the brain of endotoxin-treated rats: a pathway for the induction of non-specific symptoms of sickness? *Brain Research*, 1992. 588(2):291-296.

[47] Wan, W; Wetmore, L; Sorensen, CM; Greenberg, AH; Nance, DM. Neural and biochemical mediators of endotoxin and stress-induced c-fos expression in the rat brain. *Brain Research Bulletin*, 1994. 34(1):7-14.

[48] Bluthe, RM; Walter, V; Parnet, P; Laye, S; Lestage, J; Verrier, D; Poole, S; Stenning, BE; Kelley, KW; Dantzer, R. Lipopolysaccharide induces sickness behaviour in rats by a vagal mediated mechanism. *C R Acad Sci III*, 1994. 317(6):499-503.

[49] Nadjar, A; Bluthe, RM; May, MJ; Dantzer, R; Parnet, P. Inactivation of the cerebral NFkappaB pathway inhibits interleukin-1beta-induced sickness behavior and c-Fos expression in various brain nuclei. *Neuropsychopharmacology*, 2005. 30(8):1492-1499.

[50] Aubert, A; Vega, C; Dantzer, R; Goodall, G. Pyrogens specifically disrupt the acquisition of a task involving cognitive processing in the rat. *Brain, Behavior, and Immunity*, 1995. 9:129-148.

[51] Pugh, RC; Fleshner, M; Watkins, LR; Maier, SF; Rudy, JW. The immune system and memory consolidation: a role for the cytokine IL-1beta. *Neurosci Biobehav Rev*, 2001. 25(1):29-41.

[52] Pugh, CR; Kumagawa, K; Fleshner, M; Watkins, LR; Maier, SF; Rudy, JW. Selective effects of peripheral lipopolysaccharide administration on contextual and auditory-cue fear conditioning. *Brain, Behavior, and Immunity*, 1998. 12(3):212-229.

[53] Rudy, JW; O'Reilly, RC. Contextual fear conditioning, conjunctive representations, pattern completion, and the hippocampus. *Behav Neurosci*, 1999. 113(5):867-880.

[54] Gibertini, M; Newton, C; Friedman, H; Klein, TW. Spatial learning impairment in mice infected with Legionella pneumophila or administered exogenous interleukin-1-beta. *Brain, Behavior, and Immunity*, 1995. 9(2):113-128.

[55] Sparkman, NL; Kohman, RA; Scott, VJ; Boehm, GW. Bacterial endotoxin-induced behavioral alterations in two variations of the Morris water maze. *Physiology & Behavior*, 2005. 86(1-2):244-251.

[56] Banks, W; Farr, S; La Scola, M; Morley, J. Intravenous human interleukin-1a impairs memory processing in mice: Dependence on blood-brain barrier transport into posterior division of the septum. *The Journal of Pharmacology and Experimental Therapeutics*, 2001. 299(2):536-541.
[57] Maness, LM; Banks, WA; Zadina, JE; Kastin, AJ. Selective transport of blood-borne interleukin-1 alpha into the posterior division of the septum of the mouse brain. *Brain Research*, 1995. 700(1-2):83-88.
[58] Sparkman, NL; Buchanan, JB; Heyen, JR; Chen, J; Beverly, JL; Johnson, RW. Interleukin-6 facilitates lipopolysaccharide-induced disruption in working memory and expression of other proinflammatory cytokines in hippocampal neuronal cell layers. *The Journal of Neuroscience*, 2006. 26(42):10709-10716.
[59] Oken, RJ. Towards a unifying hypothesis of neurodegenerative diseases and a concomitant rational strategy for their prophylaxis and therapy. *Med Hypotheses*, 1995. 45(4):341-342.
[60] Toulmond, S; Parnet, P; Linthorst, AC. When cytokines get on your nerves: cytokine networks and CNS pathologies. *Trends in Neurosciences*, 1996. 19(10):409-410.
[61] Hart, S; Semple, JM. Neuropsychology and the Dementias. 1990, London, UK: Taylor and Francis.
[62] Mirra, SS; Hart, MN; Terry, RD. Making the diagnosis of Alzheimer's disease. A primer for practicing pathologists. *Arch Pathol Lab Med*, 1993. 117(2):132-144.
[63] Eikelenboom, P; Rozemuller, JM; Kraal, G; Stam, FC; McBride, PA; Bruce, ME; Fraser, H. Cerebral amyloid plaques in Alzheimer's disease but not in scrapie-affected mice are closely associated with a local inflammatory process. *Virchows Arch B Cell Pathol Incl Mol Pathol*, 1991. 60(5):329-336.
[64] Ishii, T; Haga, S. Complements, microglial cells and amyloid fibril formation. *Res Immunol*, 1992. 143(6):614-616.
[65] Rozemuller, JM; Eikelenboom, P; Pals, ST; Stam, FC. Microglial cells around amyloid plaques in Alzheimer's disease express leucocyte adhesion molecules of the LFA-1 family. *Neurosci Lett*, 1989. 101(3):288-292.
[66] Rozemuller, JM; van der Valk, P; Eikelenboom, P. Activated microglia and cerebral amyloid deposits in Alzheimer's disease. *Res Immunol*, 1992. 143(6):646-649.

[67] Weldon, DT; Rogers, SD; Ghilardi, JR; Finke, MP; Cleary, JP; O'Hare, E; Esler, WP; Maggio, JE; Mantyh, PW. Fibrillar beta-amyloid induces microglial phagocytosis, expression of inducible nitric oxide synthase, and loss of a select population of neurons in the rat CNS in vivo. *The Journal of Neuroscience*, 1998. 18(6):2161-2173.

[68] Vandenabeele, P; Fiers, W. Is amyloidogenesis during Alzheimer's disease due to an IL-1/IL-6-mediated "acute phase response" in the brain? *Immunology Today*, 1991. 12(7):217-219.

[69] McDonald, MP; Overmier, JB; Bandyopadhyay, S; Babcock, D; Cleary, J. Reversal of beta-amyloid-induced retention deficit after exposure to training and state cues. *Neurobiology of Learning and Memory*, 1996. 65(1):35-47.

[70] Yan, SD; Chen, X; Fu, J; Chen, M; Zhu, H; Roher, A; Slattery, T; Zhao, L; Nagashima, M; Morser, J; Migheli, A; Nawroth, P; Stern, D; Schmidt, AM. RAGE and amyloid-beta peptide neurotoxicity in Alzheimer's disease. *Nature*, 1996. 382(6593):685-691.

[71] Klegeris, A; Walker, DG; McGeer, PL. Interaction of Alzheimer beta-amyloid peptide with the human monocytic cell line THP-1 results in a protein kinase C-dependent secretion of tumor necrosis factor-alpha. *Brain Research*, 1997. 747(1):114-121.

[72] Klegeris, A; McGeer, PL. beta-amyloid protein enhances macrophage production of oxygen free radicals and glutamate. *Journal of Neuroscience Research*, 1997. 49(2):229-235.

[73] Houlden, H; Crook, R; Backhovens, H; Prihar, G; Baker, M; Hutton, M; Rossor, M; Martin, JJ; Van Broeckhoven, C; Hardy, J. ApoE genotype is a risk factor in nonpresenilin early-onset Alzheimer's disease families. *Am J Med Genet*, 1998. 81(1):117-121.

[74] Barger, SW; Harmon, AD. Microglial activation by Alzheimer amyloid precursor protein and modulation by apolipoprotein E. *Nature*, 1997. 388(6645):878-881.

[75] Hsiao, K; Chapman, P; Nilsen, S; Eckman, C; Harigaya, Y; Younkin, S; Yang, F; Cole, G. Correlative memory deficits, Abeta elevation, and amyloid plaques in transgenic mice. *Science*, 1996. 274(5284):99-102.

[76] Miller, B; Sarantis, M; Traynelis, SF; Attwell, D. Potentiation of NMDA receptor currents by arachidonic acid. *Nature*, 1992. 355(6362):722-725.

[77] Molina-Holgado, F; Guaza, C. Endotoxin administration induced differential neurochemical activation of the rat brain stem nuclei. *Brain Research Bulletin*, 1996. 40(3):151-156.

[78] Kuiper, MA; Visser, JJ; Bergmans, PL; Scheltens, P; Wolters, EC. Decreased cerebrospinal fluid nitrate levels in Parkinson's disease, Alzheimer's disease and multiple system atrophy patients. *J Neurol Sci*, 1994. 121(1):46-49.

[79] Rebeck, GW; Marzloff, K; Hyman, BT. The pattern of NADPH-diaphorase staining, a marker of nitric oxide synthase activity, is altered in the perforant pathway terminal zone in Alzheimer's disease. *Neurosci Lett*, 1993. 152(1-2):165-168.

[80] Milstien, S; Sakai, N; Brew, BJ; Krieger, C; Vickers, JH; Saito, K; Heyes, MP. Cerebrospinal fluid nitrite/nitrate levels in neurologic diseases. *Journal of Neurochemistry*, 1994. 63(3):1178-1180.

[81] Navarro, JA; Molina, JA; Jimenez-Jimenez, FJ; Benito-Leon, J; Orti-Pareja, M; Gasalla, T; Cabrera-Valdivia, F; Vargas, C; de Bustos, F; Arenas, J. Cerebrospinal fluid nitrate levels in patients with Alzheimer's disease. *Acta Neurologica Scandinavica*, 1996. 94(6):411-414.

[82] Medina, JH; Izquierdo, I. Retrograde messengers, long-term potentiation and memory. *Brain Res Brain Res Rev*, 1995. 21(2):185-194.

[83] Myslivecek, J; Hassmannova, J; Barcal, J; Safanda, J; Zalud, V. Inhibitory learning and memory in newborn rats influenced by nitric oxide. *Neuroscience*, 1996. 71(2):299-312.

[84] Ohno, M; Yamamoto, T; Watanabe, S. Intrahippocampal administration of the NO synthase inhibitor L-NAME prevents working memory deficits in rats exposed to transient cerebral ischemia. *Brain Research*, 1994. 634(1):173-177.

[85] Noda, Y; Yamada, K; Nabeshima, T. Role of nitric oxide in the effect of aging on spatial memory in rats. *Behavioural Brain Research*, 1997. 83(1-2):153-158.

[86] Ohno, M; Yamamoto, T; Watanabe, S. Deficits in working memory following inhibition of hippocampal nitric oxide synthesis in the rat. *Brain Research*, 1993. 632(1-2):36-40.

[87] Paakkari, I; Lindsberg, P. Nitric oxide in the central nervous system. *Ann Med*, 1995. 27(3):369-377.

[88] Fudenberg, HH; Whitten, HD; Arnaud, P; Khansari, N. Is Alzheimer's disease an immunological disorder? Observations and speculations. *Clin Immunol Immunopathol*, 1984. 32(2):127-131.

[89] Ingram, DK; Spangler, EL; Iijima, S; Ikari, H; Kuo, H; Greig, NH; London, ED. Rodent models of memory dysfunction in Alzheimer's disease and normal aging: moving beyond the cholinergic hypothesis. *Life Sci*, 1994. 55(25-26):2037-2049.

[90] Dubovik, V; Faigon, M; Feldon, J; Michaelson, DM. Decreased density of forebrain cholinergic neurons in experimental autoimmune dementia. *Neuroscience*, 1993. 56(1):75-82.
[91] Miyagawa, K; Narita, M; Akama, H; Suzuki, T. Memory impairment associated with a dysfunction of the hippocampal cholinergic system induced by prenatal and neonatal exposures to bisphenol-A. *Neurosci Lett*, 2007. 418(3):236-241.
[92] Walsh, TJ; Herzog, CD; Gandhi, C; Stackman, RW; Wiley, RG. Injection of IgG 192-saporin into the medial septum produces cholinergic hypofunction and dose-dependent working memory deficits. *Brain Research*, 1996. 726(1-2):69-79.
[93] Choi, SH; Park, CH; Koo, JW; Seo, JH; Kim, HS; Jeong, SJ; Lee, JH; Kim, SS; Suh, YH. Memory impairment and cholinergic dysfunction by centrally administered Abeta and carboxyl-terminal fragment of Alzheimer's APP in mice. *FASEB Journal*, 2001. 15(10):1816-1818.
[94] Hall, ZW; Kelly, RB. Enzymatic detachment of endplate acetylcholinesterase from muscle. *Nat New Biol*, 1971. 232(28):62-63.
[95] Alvarez, A; Opazo, C; Alarcon, R; Garrido, J; Inestrosa, NC. Acetylcholinesterase promotes the aggregation of amyloid-beta-peptide fragments by forming a complex with the growing fibrils. *Journal of Molecular Biology*, 1997. 272(3):348-361.
[96] Mesulam, MM. Alzheimer plaques and cortical cholinergic innervation. *Neuroscience*, 1986. 17(1):275-276.
[97] Moran, MA; Mufson, EJ; Gomez-Ramos, P. Co-localization of cholinesterases with beta amyloid protein in aged and Alzheimer's brains. *Acta Neuropathol*, 1993. 85(4):362-369.
[98] Mori, F; Lai, CC; Fusi, F; Giacobini, E. Cholinesterase inhibitors increase secretion of APPs in rat brain cortex. *NeuroReport*, 1995. 6(4):633-636.
[99] Inestrosa, NC; Alvarez, A; Calderon, F. Acetylcholinesterase is a senile plaque component that promotes assembly of amyloid beta-peptide into Alzheimer's filaments. *Mol Psychiatry*, 1996. 1(5):359-361.
[100] Inestrosa, NC; Alvarez, A; Perez, CA; Moreno, RD; Vicente, M; Linker, C; Casanueva, OI; Soto, C; Garrido, J. Acetylcholinesterase accelerates assembly of amyloid-beta-peptides into Alzheimer's fibrils: possible role of the peripheral site of the enzyme. *Neuron*, 1996. 16(4):881-891.
[101] Beeri, R; Andres, C; Lev-Lehman, E; Timberg, R; Huberman, T; Shani, M; Soreq, H. Transgenic expression of human acetylcholinesterase

induces progressive cognitive deterioration in mice. *Current Biology*, 1995. 5(9):1063-1071.

[102] Sihver, W; Gunther, P; Schliebs, R; Bigl, V. Repeated administration of tacrine to normal rats: effects on cholinergic, glutamatergic, and GABAergic receptor subtypes in rat brain using receptor autoradiography. *Neurochemistry International*, 1997. 31(5):693-703.

[103] Araujo, DM; Lapchak, PA; Collier, B; Quirion, R. Localization of interleukin-2 immunoreactivity and interleukin-2 receptors in the rat brain: interaction with the cholinergic system. *Brain Research*, 1989. 498(2):257-266.

[104] Hanisch, UK; Seto, D; Quirion, R. Modulation of hippocampal acetylcholine release: a potent central action of interleukin-2. *The Journal of Neuroscience*, 1993. 13(8):3368-3374.

[105] Lapchak, PA; Araujo, DM; Quirion, R; Beaudet, A. Immunoautoradiographic localization of interleukin 2-like immunoreactivity and interleukin 2 receptors (Tac antigen-like immunoreactivity) in the rat brain. *Neuroscience*, 1991. 44(1):173-184.

[106] Nistico, G; De Sarro, G. Is interleukin 2 a neuromodulator in the brain? *Trends in Neurosciences*, 1991. 14(4):146-150.

[107] Li, Y; Liu, L; Kang, J; Sheng, JG; Barger, SW; Mrak, RE; Griffin, WS. Neuronal-glial interactions mediated by interleukin-1 enhance neuronal acetylcholinesterase activity and mRNA expression. *The Journal of Neuroscience*, 2000. 20(1):149-155.

[108] Aisen, PS; Davis, KL. Inflammatory mechanisms in Alzheimer's disease: implications for therapy. *America Journal of Psychiatry*, 1994. 151(8):1105-1113.

[109] McGeer, PL; Rogers, J. Anti-inflammatory agents as a therapeutic approach to Alzheimer's disease. *Neurology*, 1992. 42(2):447-449.

[110] Rich, JB; Rasmusson, DX; Folstein, MF; Carson, KA; Kawas, C; Brandt, J. Nonsteroidal anti-inflammatory drugs in Alzheimer's disease. *Neurology*, 1995. 45(1):51-55.

[111] Rogers, J; Kirby, LC; Hempelman, SR; Berry, DL; McGeer, PL; Kaszniak, AW; Zalinski, J; Cofield, M; Mansukhani, L; Willson, P; et al. Clinical trial of indomethacin in Alzheimer's disease. *Neurology*, 1993. 43(8):1609-1611.

[112] Burger, D; Dayer, J-M, IL-1Ra, in *Cytokine Reference*, O.J.a.F. M, Editor. 2000, Academic Press: Orlando, FL. p. 319-336.

[113] Dinarello, C, IL-1a, in *Cytokine Reference*, O.J.a.F. M, Editor. 2000, Academic Press: Orlando, FL. p. 307-318.

[114] Dinarello, C, IL-1b, in *Cytokine Reference*, O.J.a.F. M, Editor. 2000, Academic Press: Orlando, FL. p. 351-374.
[115] Loddick, SA; Liu, C; Takao, T; Hashimoto, K; De Souza, EB. Interleukin-1 receptors: cloning studies and role in central nervous system disorders. *Brain Res Brain Res Rev*, 1998. 26(2-3):306-319.
[116] Vitkovic, L; Bockaert, J; Jacque, C. "Inflammatory" cytokines: neuromodulators in normal brain? *Journal of Neurochemistry*, 2000. 74(2):457-471.
[117] Berkenbosch, F; van Oers, J; del Rey, A; Tilders, F; Besedovsky, H. Corticotropin-releasing factor-producing neurons in the rat activated by interleukin-1. *Science*, 1987. 238(4826):524-526.
[118] Murray, C; McGahon, B; McBennet, S; Lynch, M. Interleukin-1b inhibits glutamate release in hippocampus of young, but not aged, rats. *Neurobiology of Aging*, 1997. 18:343-348.
[119] Plata-Salaman, CR; ffrench-Mullen, JM. Interleukin-1 beta inhibits Ca2+ channel currents in hippocampal neurons through protein kinase C. *European Journal of Pharmacology*, 1994. 266(1):1-10.
[120] Rada, P; Mark, GP; Vitek, MP; Mangano, RM; Blume, AJ; Beer, B; Hoebel, BG. Interleukin-1 beta decreases acetylcholine measured by microdialysis in the hippocampus of freely moving rats. *Brain Research*, 1991. 550(2):287-290.
[121] Cunningham, AJ; Murray, CA; O'Neill, LA; Lynch, MA; O'Connor, JJ. Interleukin-1 beta (IL-1 beta) and tumour necrosis factor (TNF) inhibit long-term potentiation in the rat dentate gyrus in vitro. *Neurosci Lett*, 1996. 203(1):17-20.
[122] Bellinger, F; Madamba, S; Siggins, G. Interleukin-1b inhibits synaptic strength and long-term potentiation in the rat CA1 hippocampus. *Brain Research*, 1993. 628:227-234.
[123] Katsuki, H; Nakai, S; Hirai, Y; Akaji, K; Kiso, Y; Satoh, M. Interleukin-1b inhibits long-term potentiation in the CA3 region of mouse hippocampal slices. *European Journal of Neuroscience*, 1990. 181:323-326.
[124] Murray, C; Lynch, M. Evidence that increased hippocampal expression of the cytokine interleukin-1b is a common trigger for age- and stress-induced impairments in long-term potentiation. *The Journal of Neuroscience*, 1998. 18(8):2974-2981.
[125] Schneider, H; Pitossi, F; Balschun, D; Wagner, A; del Rey, A; Besedovsky, HO. A neuromodulatory role of interleukin-1beta in the hippocampus. *Proc Natl Acad Sci USA*, 1998. 95(13):7778-7783.

[126] Vereker, E; O'Donnell, E; Lynch, A; Kelly, A; Nolan, Y; Lynch, M. Evidence that interleukin-1b and reactive oxygen species production play a pivotal role in stress-induced impairment of LTP in the rat dentate gyrus. *European Journal of Neuroscience*, 2001. 14:1809-1819.

[127] Oitzl, MS; van Oers, H; Schobitz, B; de Kloet, ER. Interleukin-1 beta, but not interleukin-6, impairs spatial navigation learning. *Brain Research*, 1993. 613(1):160-163.

[128] Pugh, CR; Johnson, JD; Martin, D; Rudy, JW; Maier, SF; Watkins, LR. Human immunodeficiency virus-1 coat protein gp120 impairs contextual fear conditioning: a potential role in AIDS related learning and memory impairments. *Brain Research*, 2000. 861(1):8-15.

[129] Opp, MR; Rady, PL; Hughes, TK, Jr.; Cadet, P; Tyring, SK; Smith, EM. Human immunodeficiency virus envelope glycoprotein 120 alters sleep and induces cytokine mRNA expression in rats [published errata appear in *Am J Physiol* 1996 Aug;271(2 Pt 2):section R following table of contents and 1996 Dec;271(6 Pt 3):section R following table of contents]. *American Journal of Physiology*, 1996. 270(5 Pt 2):R963-970.

[130] Matsumoto, Y; Yoshida, M; Watanabe, S; Yamamoto, T. Involvement of cholinergic and glutamatergic functions in working memory impairment induced by interleukin-1b in rats. *European Journal of Pharmacology*, 2001. 430:283-288.

[131] Barrientos, R; Higgins, E; Sprunger, D; Watkins, L; Rudy, J; Maier, S. Memory for context is impaired by a post context exposure injection of interleukin-1 beta into dorsal hippocampus. *Behavioural Brain Research*, 2002. 134:291-298.

[132] Palin, K; Bluthe, RM; Verrier, D; Tridon, V; Dantzer, R; Lestage, J. Interleukin-1beta mediates the memory impairment associated with a delayed type hypersensitivity response to bacillus Calmette-Guerin in the rat hippocampus. *Brain, Behavior, and Immunity*, 2004. 18(3):223-230.

[133] Hein, AM; Stutzman, DL; Bland, ST; Barrientos, RM; Watkins, LR; Rudy, JW; Maier, SF. Prostaglandins are necessary and sufficient to induce contextual fear learning impairments after interleukin-1 beta injections into the dorsal hippocampus. *Neuroscience*, 2007. 150(4):754-763.

[134] Yirmiya, R; Winocur, G; Goshen, I. Brain interleukin-1 is involved in spatial memory and passive avoidance conditioning. *Neurobiology of Learning and Memory*, 2002. 78(2):379-389.

[135] Konsman, J; Tridon, V; Dantzer, R. Difussion and action or intracerebroventricularly injected interleukin-1 in the CNS. *Neuroscience*, 2000. 101(4):957-967.

[136] Avital, A; Goshen, I; Kamsler, A; Segal, M; Iverfeldt, K; Richter-Levin, G; Yirmiya, R. Impaired interleukin-1 signaling is associated with deficits in hippocampal memory processes and neural plasticity. *Hippocampus*, 2003. 13(7):826-834.

[137] Labow, M; Shuster, D; Zetterstorm, M; Nunes, P; Terry, R; Cullinan, E; Bartfei, T; Solorzano, C; Moldawer, L; Chizzonite, R; McIntyre, K. Absence of IL-1 signaling and reduced inflammatory response in IL-1 type I receptor-deficient mice. *The Journal of Immunology*, 1997. 159:2452-2461.

[138] Wolf, G; Yirmiya, R; Goshen, I; Iverfeldt, K; Holmlund, L; Takeda, K; Shavit, Y. Impairment of interleukin-1 (IL-1) signaling reduces basal pain sensitivity in mice: genetic, pharmacological and developmental aspects. *Pain*, 2003. 104:471-480.

[139] Depino, AM; Alonso, M; Ferrari, C; del Rey, A; Anthony, D; Besedovsky, H; Medina, JH; Pitossi, F. Learning modulation by endogenous hippocampal IL-1: blockade of endogenous IL-1 facilitates memory formation. *Hippocampus*, 2004. 14(4):526-535.

[140] Alonso, M; Vianna, MR; Depino, AM; Mello e Souza, T; Pereira, P; Szapiro, G; Viola, H; Pitossi, F; Izquierdo, I; Medina, JH. BDNF-triggered events in the rat hippocampus are required for both short- and long-term memory formation. *Hippocampus*, 2002. 12(4):551-560.

[141] Izquierdo, I; Medina, J. Memory formation: The sequence of biochemical events in the hippocampus and its connection to activity in other brain structures. *Neurobiology of Learning and Memory*, 1997. 68:285-316.

[142] Igaz, L; Vianna, M; Medina, J; Izquierdo, I. Two time periods of hippocampal mRNA synthesis are required for memory consolidation of fear-motivated learning. *The Journal of Neuroscience*, 2002. 22(15):6781-6789.

[143] Nguyen, K; Deak, T; Owens, S; Kohno, T; Fleshner, M; Watkins, L; Maier, S. Exposure to acute stress induces brain interleukin-1b protein in rat. *The Journal of Neuroscience*, 1998. 18(6):2239-2246.

[144] Bianchi, M; Dib, B; Panerai, A. Interleukin-1 and nociception in the rat. *Journal of Neuroscience Research*, 1998. 53:645-650.

[145] Zou, CJ; Liu, JD; Zhou, YC. Roles of central interleukin-1 on stress-induced-hypertension and footshock-induced-analgesia in rats. *Neurosci Lett*, 2001. 311(1):41-44.
[146] Pittenger, C; Huang, YY; Paletzki, RF; Bourtchouladze, R; Scanlin, H; Vronskaya, S; Kandel, ER. Reversible inhibition of CREB/ATF transcription factors in region CA1 of the dorsal hippocampus disrupts hippocampus-dependent spatial memory. *Neuron*, 2002. 34:447-462.
[147] Shimizu, E; Tang, Y-P; Rampon, C; Tsien, J. NMDA receptor-dependent synaptic reinforcement as a crucial process for memory consolidation. *Science*, 2000. 290:1170-1174.
[148] Genoux, D; Haditsch, U; Knobloch, M; Michalon, A; Storm, D; Mansuy, I. Protein phosphatase 1 is a molecular constraint on learning and memory. *Nature*, 2002. 418:970-975.
[149] Acquas, E; Wilson, C; Fibiger, H. Conditioned and unconditioned stimuli increase frontal cortical and hippocampal acetylcholine release: Effects of novelty, habituation, and fear. *The Journal of Neuroscience*, 1996. 16(9):3089-3096.
[150] Thiel, C; Huston, J; Schwarting, R. Hippocampal acetylcholine and habituation learning. *Neuroscience*, 1998. 85(4):1253-1262.
[151] Vianna, M; Alonso, M; Viola, H; Quevedo, J; de Paris, F; Furman, M; Levi de Stein, M; Medina, J; Izquierdo, I. Role of hippocampal signaling pathways in long-term memory formation of a nonassociative learning task in the rat. *Learning & Memory*, 2000. 7:333-340.
[152] Izquierdo, I; Medina, J. The biochemistry of memory formation and its regulation by hormones and neurotransmitters. *Psychobiology*, 1997. 25:1-9.
[153] McGaugh, J; Roozendaal, B. Role of adrenal stress hormones in forming lasting memories in the brain. *Current Opinion in Neurobiology*, 2002. 12:205-210.
[154] Goshen, I; Kreisel, T; Ounallah-Saad, H; Renbaum, P; Zalzstein, Y; Ben-Hur, T; Levy-Lahad, E; Yirmiya, R. A dual role for interleukin-1 in hippocampal-dependent memory processes. *Psychoneuroendocrinology*, 2007. 32(8-10):1106-1115.
[155] Spulber, S; Mateos, L; Oprica, M; Cedazo-Minguez, A; Bartfai, T; Winblad, B; Schultzberg, M. Impaired long-term memory consolidation in transgenic mice overexpressing the human soluble form of IL-1ra in the brain. *Journal of Neuroimmunology*, 2009. 208(1-2):46-53.
[156] Trompet, S; de Craen, AJ; Slagboom, P; Shepherd, J; Blauw, GJ; Murphy, MB; Bollen, EL; Buckley, BM; Ford, I; Gaw, A; Macfarlane,

PW; Packard, CJ; Stott, DJ; Jukema, JW; Westendorp, RG. Genetic variation in the interleukin-1 beta-converting enzyme associates with cognitive function. The PROSPER study. *Brain*, 2008. 131(Pt 4):1069-1077.
[157] Tsai, SJ; Hong, CJ; Liu, ME; Hou, SJ; Yen, FC; Hsieh, CH; Liou, YJ. Interleukin-1 beta (C-511T) genetic polymorphism is associated with cognitive performance in elderly males without dementia. *Neurobiology of Aging*, 2008.
[158] Pociot, F; Molvig, J; Wogensen, L; Worsaae, H; Nerup, J. A TaqI polymorphism in the human interleukin-1 beta (IL-1 beta) gene correlates with IL-1 beta secretion in vitro. *Eur J Clin Invest*, 1992. 22(6):396-402.
[159] Hirano, T; Akira, S; Taga, T; Kishimoto, T. Biological and clinical aspects of interleukin 6. *Immunology Today*, 1990. 11(12):443-449.
[160] Pousset, F. Developmental expression of cytokine genesin the cortex and hippocampus of the rat central nervous system. *Developmental Brain Research*, 1994. 81:143-146.
[161] Gadient, RA; Otten, U. Expression of interleukin-6 (IL-6) and interleukin-6 receptor (IL-6R) mRNAs in rat brain during postnatal development. *Brain Research*, 1994. 637:10-14.
[162] Schöbitz, B; de Kloet, ER; Sutanto, W; Holsboer, F. Cellular localization of interleukin 6 mRNA and interleukin 6 receptor mRNA in rat brain. *European Journal of Neuroscience*, 1993. 5:1426-1435.
[163] Sallmann, S; Juttler, E; Prinz, S; Petersen, N; Knopf, U; Weiser, T; Schwaninger, M. Induction of interleukin-6 by depolarization of neurons. *The Journal of Neuroscience*, 2000. 20(23):8637-8642.
[164] Kushima, Y; Hama, T; Hatanaka, H. Interleukin-6 as a neurotrophic factor for promoting the survival of cultured catecholaminergic neurons in a chemically defined medium from fetal and postnatal rat midbrains. *Neurosci Res*, 1992. 13(4):267-280.
[165] Kushima, Y; Hatanaka, H. Interleukin-6 and leukemia inhibitory factor promote the survival of acetylcholinesterase-positive neurons in culture from embryonic rat spinal cord. *Neurosci Lett*, 1992. 143(1-2):110-114.
[166] Hama, T; Kushima, Y; Miyamoto, M; Kubota, M; Takei, N; Hatanaka, H. Interleukin-6 improves the survival of mesencephalic catecholaminergic and septal cholinergic neurons from postnatal, two-week-old rats in cultures. *Neuroscience*, 1991. 40(2):445-452.

[167] Satoh, T; Nakamura, S; Taga, T; Matsuda, T; Hirano, T; Kishimoto, T; Kaziro, Y. Induction of neuronal differentiation in PC12 cells by B-cell stimulatory factor 2/interleukin 6. *Mol Cell Biol*, 1988. 8(8):3546-3549.
[168] Plata-Salaman, CR. Food intake suppression by growth factors and platelet peptides by direct action in the central nervous system. *Neurosci Lett*, 1988. 94(1-2):161-166.
[169] LeMay, LG; Vander, AJ; Kluger, MJ. Role of interleukin 6 in fever in rats. *American Journal of Physiology*, 1990. 258(3 Pt 2):R798-803.
[170] Naitoh, Y; Fukata, J; Tominaga, T; Nakai, Y; Tamai, S; Mori, K; Imura, H. Interleukin-6 stimulates the secretion of adrenocorticotropic hormone in conscious, freely-moving rats. *Biochem Biophys Res Commun*, 1988. 155(3):1459-1463.
[171] Campbell, IL; Abraham, CR; Masliah, E; Kemper, P; Inglis, JD; Oldstone, MBA; Mucke, L. Neurologic disease induced in transgenic mice by cerebral overexpression of interleukin 6. *Proceedings of National Academy of Sciences*, 1993. 90:10061-10065.
[172] Heyser, CJ; Masliah, E; Samimi, A; Campbell, IL; Gold, LH. Progressive decline in avoidance learning paralleled by inflammatory neurodegeneration in transgenic mice expressing interleukin 6 in the brain. *Proc Natl Acad Sci USA*, 1997. 94(4):1500-1505.
[173] Braida, D; Sacerdote, P; Panerai, AE; Bianchi, M; Aloisi, AM; Iosue, S; Sala, M. Cognitive function in young and adult IL (interleukin)-6 deficient mice. *Behavioural Brain Research*, 2004. 153(2):423-429.
[174] Baier, PC; May, U; Scheller, J; Rose-John, S; Schiffelholz, T. Impaired hippocampus-dependent and -independent learning in IL-6 deficient mice. *Behavioural Brain Research*, 2009. 200(1):192-196.
[175] Weaver, JD; Huang, MH; Albert, M; Harris, T; Rowe, JW; Seeman, TE. Interleukin-6 and risk of cognitive decline: MacArthur studies of successful aging. *Neurology*, 2002. 59(3):371-378.
[176] Harding, DR; Humphries, SE; Whitelaw, A; Marlow, N; Montgomery, HE. Cognitive outcome and cyclo-oxygenase-2 gene (-765 G/C) variation in the preterm infant. *Arch Dis Child Fetal Neonatal Ed*, 2007. 92(2):F108-112.
[177] Mattson, MP; Bruce, AJ; Blane, EM, Cellular actions of tumor necrosis factor in ischemic brain injury, in *Pharmacology of Cerebral Ischemia.*, J.Krieglstein, Editor. 1996. p. 93-105.
[178] Munoz-Fernandez, MA; Fresno, M. The role of tumour necrosis factor, interleukin 6, interferon-gamma and inducible nitric oxide synthase in

the development and pathology of the nervous system. *Progress in Neurobiology*, 1998. 56(3):307-340.
[179] Chen, G; Goeddel, DV. TNF-R1 signaling: a beautiful pathway. *Science*, 2002. 296(5573):1634-1635.
[180] Baud, V; Karin, M. Signal transduction by tumor necrosis factor and its relatives. *Trends Cell Biol*, 2001. 11(9):372-377.
[181] MacEwan, DJ. TNF receptor subtype signalling: differences and cellular consequences. *Cell Signal*, 2002. 14(6):477-492.
[182] Blond, D; Campbell, SJ; Butchart, AG; Perry, VH; Anthony, DC. Differential induction of interleukin-1beta and tumour necrosis factor-alpha may account for specific patterns of leukocyte recruitment in the brain. *Brain Research*, 2002. 958(1):89-99.
[183] Depino, A; Ferrari, C; Pott Godoy, MC; Tarelli, R; Pitossi, FJ. Differential effects of interleukin-1beta on neurotoxicity, cytokine induction and glial reaction in specific brain regions. *Journal of Neuroimmunology*, 2005. 168(1-2):96-110.
[184] Probert, L; Akassoglou, K; Pasparakis, M; Kontogeorgos, G; Kollias, G. Spontaneous inflammatory demyelinating disease in transgenic mice showing central nervous system-specific expression of tumor necrosis factor alpha. *Proc Natl Acad Sci USA*, 1995. 92(24):11294-11298.
[185] Bruce, AJ; Boling, W; Kindy, MS; Peschon, J; Kraemer, PJ; Carpenter, MK; Holtsberg, FW; Mattson, MP. Altered neuronal and microglial responses to excitotoxic and ischemic brain injury in mice lacking TNF receptors. *Nat Med*, 1996. 2(7):788-794.
[186] Liu, J; Marino, MW; Wong, G; Grail, D; Dunn, A; Bettadapura, J; Slavin, AJ; Old, L; Bernard, CC. TNF is a potent anti-inflammatory cytokine in autoimmune-mediated demyelination. *Nat Med*, 1998. 4(1):78-83.
[187] Butler, MP; O'Connor, JJ; Moynagh, PN. Dissection of tumor-necrosis factor-alpha inhibition of long-term potentiation (LTP) reveals a p38 mitogen-activated protein kinase-dependent mechanism which maps to early-but not late-phase LTP. *Neuroscience*, 2004. 124(2):319-326.
[188] Cumiskey, D; Butler, MP; Moynagh, PN; O'Connor J, J. Evidence for a role for the group I metabotropic glutamate receptor in the inhibitory effect of tumor necrosis factor-alpha on long-term potentiation. *Brain Research*, 2007. 1136(1):13-19.
[189] Beattie, EC; Stellwagen, D; Morishita, W; Bresnahan, JC; Ha, BK; Von Zastrow, M; Beattie, MS; Malenka, RC. Control of synaptic strength by glial TNFalpha. *Science*, 2002. 295(5563):2282-2285.

[190] Stellwagen, D; Malenka, RC. Synaptic scaling mediated by glial TNF-alpha. *Nature*, 2006. 440(7087):1054-1059.

[191] Fiore, M; Probert, L; Kollias, G; Akassoglou, K; Alleva, E; Aloe, L. Neurobehavioral alterations in developing transgenic mice expressing TNF-alpha in the brain. *Brain, Behavior, and Immunity*, 1996. 10(2):126-138.

[192] Aloe, L; Properzi, F; Probert, L; Akassoglou, K; Kassiotis, G; Micera, A; Fiore, M. Learning abilities, NGF and BDNF brain levels in two lines of TNF-alpha transgenic mice, one characterized by neurological disorders, the other phenotypically normal. *Brain Research*, 1999. 840(1-2):125-137.

[193] Fiore, M; Angelucci, F; Alleva, E; Branchi, I; Probert, L; Aloe, L. Learning performances, brain NGF distribution and NPY levels in transgenic mice expressing TNF-alpha. *Behavioural Brain Research*, 2000. 112(1-2):165-175.

[194] Iida, R; Saito, K; Yamada, K; Basile, AS; Sekikawa, K; Takemura, M; Fujii, H; Wada, H; Seishima, M; Nabeshima, T. Suppression of neurocognitive damage in LP-BM5-infected mice with a targeted deletion of the TNF-alpha gene. *FASEB Journal*, 2000. 14(7):1023-1031.

[195] McAfoose, J; Koerner, H; Baune, BT. The effects of TNF deficiency on age-related cognitive performance. *Psychoneuroendocrinology*, 2009. 34(4):615-619.

[196] Baune, BT; Ponath, G; Rothermundt, M; Riess, O; Funke, H; Berger, K. Association between genetic variants of IL-1beta, IL-6 and TNF-alpha cytokines and cognitive performance in the elderly general population of the MEMO-study. *Psychoneuroendocrinology*, 2008. 33(1):68-76.

INDEX

A

acetylcholine, 13, 15, 17, 18, 40, 41, 44
acetylcholinesterase, 15, 39, 40, 45
Ach, 13, 15
acid, 3, 8, 37
ACTH, 17, 29
activation, ix, 4, 9, 10, 11, 12, 13, 14, 16, 23, 25, 37
active transport, 4
acute, ix, 10, 22, 37, 43
acute stress, 43
adenoviral vectors, 22
adhesion, 3, 4, 31, 36
administration, 4, 8, 10, 14, 17, 18, 19, 32, 33, 34, 35, 37, 38, 40
adrenocorticotropic hormone, 17, 29, 46
adult, 8, 22, 23, 26, 46
age, 13, 15, 16, 23, 26, 41, 48
agent, 15
agents, 8, 40
aggregates, 12
aggregation, 39
aging, 38, 46
agonist, 18
AIDS, 3, 12, 42
allele, 13, 24, 26
alpha, 2, 29, 32, 34, 36, 37, 47, 48
alters, 42

amino, 3
amino acid, 3
amygdala, 9
amyloid, 12, 13, 15, 16, 27, 29, 36, 37, 39
amyloid beta, 39
amyloid deposits, 36
amyloid fibril formation, 36
amyloid fibrils, 15
amyloid plaques, 12, 15, 27, 36, 37
amyloid precursor protein, 13, 15, 29, 37
amyloidosis, 13
analgesia, 25, 44
angiogenesis, 23
animal models, ix, 15
animals, ix, 7, 11, 14, 15, 19, 20, 21, 26, 33
anorexia, 11, 19
antagonism, 18
antagonist, 9, 17, 20, 34
antigen, 40
anti-inflammatory drugs, 16, 40
apoptosis, 25
APP, 13, 15, 16, 29, 39
appetite, 7
arachidonic acid, 37
astrocyte, 3
astrocytes, 1, 3, 12, 13, 23, 25
ataxia, 23
ATF, 44
atrophy, 12, 38

autocrine, 1
autoimmune, 2, 39, 47
autoimmune disease, 2
autoimmune diseases, 2
autoradiography, 40
avoidance, 11, 19, 20, 21, 23, 25, 42, 46

B

bacillus, 19, 42
bacillus Calmette-Guerin, 19, 42
bacteria, 4, 10
bacterial, 2, 3, 7
bacterial infection, 7
barrier, 4, 11, 23, 31, 32, 36
basal forebrain, 15
BBB, 11
B-cell, 46
BCG, 19
BDNF, 19, 21, 26, 43, 48
behavior, ix, 7, 8, 9, 11, 13, 15, 19, 24, 32, 33, 34, 35
behavioral change, 24
behavioral disorders, 23
behavioral effects, 4, 9, 14, 15, 34
beneficial effect, ix, 10
binding, 13, 17, 25
biochemistry, 44
bisphenol, 39
blocks, 4, 18, 22, 25, 32, 33
blood, 4, 11, 17, 23, 31, 32, 36
blood-brain barrier, 4, 11, 23, 32, 36
body temperature, 8, 18
body weight, 8, 25
brain, ix, 1, 3, 4, 5, 7, 8, 9, 11, 12, 13, 14, 15, 16, 17, 18, 19, 20, 21, 22, 23, 24, 25, 26, 27, 31, 32, 33, 34, 35, 36, 37, 39, 40, 41, 43, 44, 45, 46, 47, 48
brain damage, 3
brain development, 24, 31
brain injury, 25
brain stem, 37
brain structure, 43

C

Ca^{2+}, 13, 14, 41
calcium, 17
cAMP, 14
carboxyl, 15, 39
caspases, 25
Catecholamines, 31
cats, 32
cell, 3, 4, 13, 17, 23, 31, 36, 37, 46
cell adhesion, 3
cell line, 13, 37
cell surface, 31
central nervous system, ix, 1, 12, 25, 29, 32, 33, 35, 38, 41, 45, 46, 47
cerebellum, 23
cerebral cortex, 15
cerebral ischemia, 14, 38
cerebrospinal fluid, 14, 38
c-fos, 9, 35
chemokines, 31
children, 24
chloride, 34
cholinergic, 15, 38, 39, 40, 42, 45
cholinergic neurons, 15, 39, 45
cholinesterase, 18
circulation, 1, 4
citotoxic, 24
classes, 23
cloning, 41
CNS, 1, 2, 3, 11, 14, 15, 23, 25, 29, 31, 32, 36, 37, 43
coding, 23
cognition, 12
cognitive deficit, 9, 13, 14
cognitive deficits, 9, 13, 14
cognitive development, 24
cognitive dysfunction, 13, 15
cognitive function, 9, 15, 16, 22, 24, 45
cognitive impairment, 2, 11, 12, 15, 19, 27
cognitive performance, 22, 45, 48
cognitive process, 13, 32, 35
cognitive processing, 35
cognitive test, 22

communication, ix, 7, 11, 24, 32, 34
complement, 3, 13
components, 7
concentration, 4, 15
conception, 22
conditioned response, 10
conditioning, 10, 18, 19, 20, 21, 35, 42
Congress, vi
consensus, 22
consolidation, ix, 18, 19, 20, 21, 22, 35, 43, 44
consumption, 7
control, 10, 16, 20, 23
cortex, 12, 16, 23, 39, 45
corticosterone, 4, 32
CREB, 44
cross-sectional, 2
cues, 37
culture, 23, 45
cytokine, ix, 1, 2, 4, 5, 8, 11, 14, 15, 16, 17, 20, 21, 22, 23, 24, 25, 26, 27, 31, 32, 33, 34, 35, 36, 41, 42, 45, 47
cytokine networks, 32, 36
cytokine receptor, 1, 32
cytokines, ix, 1, 2, 3, 4, 5, 7, 8, 9, 11, 12, 13, 14, 16, 18, 27, 31, 32, 36, 41, 48
cytotoxic, 3

D

de novo, 20
death, 14
defense, 22
defense mechanisms, 22
deficiency, 26, 48
deficit, 10, 14, 18, 37
deficits, ix, 9, 11, 13, 14, 15, 22, 23, 25, 26, 38, 39, 43
degenerative disease, 12
delivery, 10
dementia, 2, 3, 12, 39, 45
demyelinating disease, 47
demyelination, 25, 47
density, 39
dentate gyrus, 23, 41, 42

depolarization, 23, 45
deposits, 36
depressed, 9
depression, 2, 7
detachment, 39
developing brain, 3
deviation, 21
differentiation, 22, 46
direct action, 46
disorder, 12, 38
distribution, 48
division, 1, 11, 36
dopamine, 14
drugs, 16, 40
duration, 16
dysregulation, 14, 15

E

elderly, 22, 24, 26, 45, 48
embryogenesis, 1
endothelial cell, 4, 31
endothelial cells, 4, 31
energy, 7, 27
energy consumption, 7
engagement, 8
enlargement, 12
environment, 21
epileptic seizures, 23
ERK1, 21
etiology, 12
excitotoxic, 47
exposure, 10, 18, 19, 25, 31, 37, 42, 43

F

family, 1, 34, 36
fatigue, 7
fear, 8, 10, 18, 19, 20, 21, 35, 42, 43, 44
feedback, 17, 31
feeding, 8, 34
fetal, 45
fever, 5, 7, 8, 10, 23, 46
fibrils, 39

fluid, 14, 38
food, 8, 9, 10, 18, 23, 33, 34
food intake, 8, 9, 23
Ford, 44
forebrain, 39
forgetting, 26
free radical, 3, 37
free radicals, 3, 37
freezing, 19
fungal, 2

G

GABA, 3
GABAergic, 40
gene, 9, 13, 21, 22, 23, 24, 33, 45, 46, 48
gene expression, 21, 33
genotype, 22, 37
genotypes, 24
GFAP, 21, 22, 23, 29
glia, 5, 16
glial, 3, 5, 12, 13, 14, 17, 23, 29, 40, 47, 48
glial cells, 3, 5, 12, 13, 14, 17
glial fibrillary acidic protein, 23, 29
gliosis, 3, 27
glucocorticoids, 17, 20
glutamate, 3, 13, 17, 18, 25, 37, 41, 47
glutamatergic, 40, 42
glycoprotein, 42
Gram-negative, 4
gram-negative bacteria, 10
Gram-positive, 4
groups, 10, 20
growth, 1, 22, 23, 46
growth factor, 1, 46
growth factors, 1, 46

H

habituation, 21, 44
handling, 20
health, 11, 27
hematopoiesis, 22

hippocampal, ix, 10, 11, 16, 18, 19, 20, 21, 23, 36, 38, 39, 40, 41, 43, 44
hippocampus, 8, 9, 10, 11, 14, 15, 16, 18, 19, 20, 21, 22, 23, 26, 35, 41, 42, 43, 44, 45, 46
Hippocampus, 43
HIV, 18
homeostasis, 8
hormone, 17, 29, 46
hormones, 21, 31, 44
host, ix, 3, 7, 22
HPA, 17, 23, 29
human, ix, 11, 13, 15, 23, 36, 37, 39, 44, 45
humans, ix, 7, 15, 22
hydro, 4
hydrolysis, 15
hydrophilic, 4
hyperactivity, 15
hyperalgesia, 8, 34
hypersensitivity, 42
hypertension, 44
hypothalamic, 9, 17, 23
hypothalamus, 1, 4, 9, 17, 29
hypothalamus-pituitary-adrenal (HPA), 1, 29
hypothesis, 12, 15, 19, 36, 38

I

IFN, 9
IgG, 39
IL-1, ix, 1, 2, 4, 7, 8, 9, 10, 11, 13, 16, 17, 18, 19, 20, 21, 22, 24, 25, 27, 29, 32, 33, 35, 37, 40, 41, 43, 44, 45, 48
IL-2, 15, 25
IL-6, ix, 2, 4, 7, 8, 9, 11, 16, 18, 22, 23, 24, 27, 29, 37, 45, 46, 48
immune activation, ix, 10
immune cells, 10
immune response, 4, 12, 22
immune system, ix, 1, 7, 24, 33, 35
immunity, ix, 3, 33
immunodeficiency, 42
immunological, 38

Index

immunopathogenesis, 16
immunoreactivity, 40
immunoregulation, ix
impairments, 10, 11, 12, 13, 14, 20, 25, 41, 42
in vitro, 13, 14, 16, 41, 45
in vivo, 18, 23, 37
indomethacin, 16, 40
induction, 4, 14, 16, 18, 19, 33, 34, 35, 47
infection, ix, 2, 7, 10, 25, 27, 31
infectious, 8
inflammation, 2, 5, 8, 9, 10, 11, 12, 13, 14, 15, 16, 19, 23, 25, 27, 33
inflammatory, ix, 1, 2, 3, 4, 7, 9, 11, 12, 13, 16, 17, 23, 27, 31, 36, 40, 43, 46, 47
inflammatory mediators, 16
inflammatory response, 7, 17, 43
inflammatory responses, 17
inhibition, 15, 38, 44, 47
inhibitor, 14, 18, 38
inhibitors, 16, 39
inhibitory, 20, 21, 22, 45, 47
inhibitory effect, 47
injection, 8, 9, 10, 14, 15, 18, 19, 32, 42
injections, 10, 42
injury, vi, 3, 16
innervation, 39
iNOS, 14
interaction, 4, 40
interactions, 1, 16, 32, 40
interferon, 46
interferons, 7, 25
interleukin, 1, 2, 7, 17, 22, 29, 31, 32, 33, 34, 35, 36, 40, 41, 42, 43, 44, 45, 46, 47
interleukin-1, 2, 7, 17, 31, 32, 33, 34, 35, 36, 40, 41, 42, 43, 44, 45, 47
interleukin-2, 32, 40
interleukin-6, 2, 7, 34, 42, 45
intracranial, 14
intraperitoneal, 4, 9, 10, 29, 32
intravenous, 29, 32
intravenously, 11

ischemia, 14, 38
ischemic, 14, 46, 47
ischemic brain injury, 46, 47

J

jumping, 20
juveniles, 8

K

Kinase, ii
knockout, 19, 22, 29

L

lamina, 4
latency, 20, 25
leakage, 4
learning, ix, 2, 10, 11, 12, 14, 18, 19, 20, 21, 22, 23, 24, 25, 27, 35, 38, 42, 43, 44, 46
learning task, 20, 21, 23, 44
Legionella, 10, 35
Legionella pneumophila, 10, 35
lethargy, 7
leucocyte, 36
leukemia, 45
leukocyte, 47
LFA, 36
ligands, 17
lipopolysaccharide, 29, 31, 33, 34, 35, 36
Lipoprotein, ii, iii
lithium, 34
localization, 12, 39, 40, 45
locomotor activity, 8, 20
London, 36, 38
long-term memory, 19, 20, 29, 43, 44
long-term potentiation, ix, 18, 29, 38, 41, 47
LPS, 4, 8, 9, 10, 11, 29, 33
LTP, 18, 25, 29, 42, 47
lupus, 2

lymphocyte, 31

M

macrophage, 37
macrophages, 3, 4, 8, 24, 25, 31, 35
magnetic, vi
malaise, 7
males, 22, 45
MAPK, 25
maze tasks, 15
measures, 14, 15, 16
mediators, 2, 4, 16, 35
memory, ix, 2, 10, 11, 12, 14, 15, 16, 18, 19, 20, 21, 22, 23, 24, 26, 27, 29, 35, 36, 37, 38, 39, 42, 43, 44
memory deficits, 12, 18, 37
memory formation, 2, 11, 19, 20, 21, 22, 24, 26, 43, 44
memory performance, 14, 15, 18
memory processes, ix, 19, 21, 24, 43, 44
messengers, 1, 32, 38
metabolic, ix, 7, 14
metabolism, 25
metabotropic glutamate receptor, 25, 47
metabotropic glutamate receptors, 25
mice, 5, 8, 10, 11, 13, 14, 15, 17, 19, 21, 22, 23, 24, 25, 26, 31, 33, 34, 35, 36, 37, 39, 40, 43, 44, 46, 47, 48
microdialysis, 41
microglia, 1, 3, 12, 13, 16, 23, 25, 35, 36
microglial, 8, 12, 13, 36, 37, 47
microglial cells, 8, 12, 36
microvasculature, 9
mitogen, 47
mitogen-activated protein kinase, 47
models, ix, 15, 18, 38
modulation, ix, 4, 11, 16, 19, 21, 37, 43
molecular mechanisms, 19
molecular weight, 4
molecules, ix, 1, 3, 4, 5, 17, 31, 36
monocytes, 4
morphology, 4
motor activity, 23
motor function, 23

mouse, 31, 36, 41
mRNA, 18, 20, 23, 33, 34, 40, 42, 43, 45
multiple sclerosis, 2
muscle, 39

N

National Academy of Sciences, 46
natural, 8, 20
necrosis, ix, 2, 7, 29, 34, 37, 41, 46, 47
neonatal, 39
nerve, 1, 4, 9
nerve growth factor, 1
nerves, 4, 32, 36
nervous system, 1, 31, 47
network, 17
neuritis, 15
neurocognitive damage, 48
neurodegeneration, 23, 25, 46
neurodegenerative, 2, 12, 36
neurodegenerative disease, 2, 12, 36
neurodegenerative diseases, 2, 12, 36
neuroendocrine, ix, 4, 7
neurologic symptom, 25
neurological disorder, 3, 48
neuromodulator, 16, 40
neurons, 1, 3, 4, 5, 14, 15, 17, 23, 25, 37, 39, 41, 45
neuropathological, 27
neuropathology, 12, 13
neuroprotective, 2, 25
neuropsychiatric disorders, 2
neurotoxic, 5, 12, 16, 25
neurotoxicity, 12, 23, 37, 47
neurotransmission, 14, 15
neurotransmitter, 13, 16, 25
neurotransmitters, 3, 21, 44
neurotrophic, 19, 26, 45
neurotrophic factors, 26
New York, v, vi
Newton, 35
nitrate, 14, 38
nitric oxide (NO), 3, 14, 29, 37, 38, 46
nitric oxide synthase, 37, 38, 46
nitrogen, 3, 32

NMDA, 18, 37, 44
NO synthase, 14, 38
nociception, 43
nociceptive, 20, 22
non-steroidal anti-inflammatory drugs, 16
normal, 3, 10, 11, 15, 17, 20, 23, 27, 38, 40, 41, 48
normal aging, 38
NOS, 14, 15
novelty, 44
nuclei, 9, 35, 37
nucleus, 9, 17
nucleus tractus solitarius, 9

O

object recognition, 24
organ, 4
organism, 7
oxide, 13, 38
oxygen, 3, 32, 37, 42

P

pain, 8, 20, 34, 43
paracrine, 1, 4
paraventricular, 9, 17
paraventricular nucleus, 17
parenchyma, 4
parenchymal, 25
parenchymal infiltration, 25
Paris, 44
Parkinson, 2, 12, 38
passive, 19, 25, 42
pathogenesis, 12, 31
pathogenic, 12, 13, 15
pathogens, 7
pathology, 15, 16, 33, 47
pathophysiology, 12
pathways, 8, 11, 14, 21
patients, 12, 13, 14, 15, 16, 38
PC12 cells, 46
peptide, 9, 12, 13, 15, 37, 39

peptides, 39, 46
periodic, 16
Peripheral, 10, 34
peripheral nervous system, 2
peritoneal, 31
permeability, 23, 32
phagocytic, 12, 13
phagocytosis, 2, 37
pharmacological, 43
pharmacology, 33
phenotype, 26
physiological, ix, 2, 3, 7, 15, 18, 21, 22, 25
physiology, 3, 8, 33
pituitary, 17, 23, 34, 35
placebo, 16
plaque, 13, 39
plaques, 12, 13, 39
plasma, 24, 26
plasma levels, 24, 26
plasticity, 20, 43
platelet, 46
play, ix, 1, 15, 19, 42
polymorphism, 22, 26, 45
polymorphisms, 22
polypeptides, 1
population, 2, 3, 37, 48
press, 10
prion diseases, 2
probability, 10
production, 4, 5, 13, 20, 23, 25, 27, 37, 42
proinflammatory, 18, 21, 36
pro-inflammatory, ix, 1, 2, 3, 4, 7, 9, 27
proliferation, 3
promoter, 21, 22, 23
promoter region, 22
prophylaxis, 36
prostaglandin, 19
prostaglandins, 19, 32, 34
prostanoids, 3
protease inhibitors, 16
proteases, 16
protein, 13, 17, 18, 23, 24, 37, 39, 41, 42, 43

protein kinase C, 17, 37, 41
proteins, 3, 4, 7, 16, 19
Proteins, ii
pyramidal, 23

R

RAGE, 37
rain, 27
range, 3, 21, 22
rat, 10, 15, 23, 32, 34, 35, 37, 38, 39, 40, 41, 42, 43, 44, 45
rats, 4, 8, 9, 10, 14, 17, 18, 19, 33, 35, 38, 40, 41, 42, 44, 45, 46
reactive nitrogen, 32
reactive oxygen, 3, 42
reactive oxygen species, 42
receptors, 3, 4, 5, 9, 15, 17, 24, 25, 32, 33, 40, 41, 47
recognition, 24, 31
recognition test, 24
refractory, 11, 20
regulation, ix, 20, 27, 31, 44
reinforcement, 44
relationship, 14, 24
relationships, 33
relatives, 47
remodeling, 1
repair, 1
retardation, 25
retention, 11, 25, 37
reversal learning, 24
risk, 13, 14, 24, 37, 46
RNA, 8, 31
rodent, 8, 13
rodents, 4, 8, 34
runaway, 18

S

saline, 10, 15, 18
scaling, 48
schizophrenia, 2, 31
secrete, 13
secretion, ix, 3, 37, 39, 45, 46
seizure, 23
senile, 12, 39
senile plaques, 12
sensitivity, 21, 43
septum, 11, 36, 39
serotonin, 13
services, vi
severity, 14
sexual activity, 7
shock, 20, 21
short-term, 20, 29
short-term memory, 20, 29
signal transduction, 13
signaling, ix, 1, 3, 9, 19, 22, 24, 31, 43, 44, 47
signaling pathway, 9, 44
signaling pathways, 44
signalling, 21, 22, 47
signals, 3, 4, 5, 7, 9
sites, 17
sleep, 42
social withdrawal, 9
spatial, 10, 11, 13, 14, 15, 18, 19, 23, 24, 26, 38, 42, 44
spatial learning, 13, 18, 23, 24
spatial memory, 19, 26, 38, 42, 44
species, 3, 8, 22, 32, 34, 42
specificity, 25
speed, 20, 25, 26
spinal cord, 45
stages, 25
stimulus, 9, 10, 21
STM, 20, 29
strength, 41, 47
stress, 20, 24, 35, 41, 42, 43, 44
stressors, 8
striatum, 8, 16, 23
stroke, 2
substances, 8
successful aging, 46
suppression, 8, 34, 46
suppressors, 21
survival, 23, 45
susceptibility, 24

symptoms, ix, 11, 25, 35
synapse, 25
synaptic strength, 25, 41, 47
synaptic transmission, 15
syndrome, ix
Synergy, 34
synthesis, 3, 4, 8, 9, 13, 14, 38, 43
systemic lupus erythematosus, 2

T

targets, 2, 4
temperature, 8, 18
temporal, 23
thalamus, 8
therapeutic interventions, 2
therapy, 36, 40
thresholds, 20
time periods, 43
tissue, 1, 3
TLR2, 4
TLR4, 4
T-maze, 11, 13
TNF, ix, 2, 4, 7, 8, 9, 13, 16, 24, 25, 26, 27, 29, 32, 41, 47, 48
TNF-alpha, 29, 32, 48
toll-like, 4, 31
training, 10, 11, 18, 19, 20, 37
transcription, 44
transcription factor, 44
transcription factors, 44
transcripts, 34, 35
transduction, 14, 47
transgenic, 13, 15, 22, 23, 25, 37, 39, 44, 46, 47, 48
transgenic mice, 15, 22, 25, 37, 44, 46, 47, 48
transgenic mouse, 13, 23
transmission, 9, 15
transport, 4, 11, 36
trauma, 1, 2, 12
tremor, 23

trial, 18, 40
tumor, ix, 2, 7, 29, 34, 37, 46, 47
tumor necrosis factor, ix, 2, 7, 29, 34, 37, 46, 47
tumour, 41, 46, 47

U

unconditioned, 44
underlying mechanisms, 21

V

vagus, 4, 9, 32
vagus nerve, 4, 9, 32
variation, 10, 45, 46
vasculature, 4
vector, 20
ventricle, 9, 18
ventricles, 11, 12, 18, 19
virus, 10, 42

W

water, 10, 11, 13, 14, 15, 18, 19, 20, 29, 35
water maze, 10, 11, 13, 14, 15, 18, 19, 20, 29, 35
weakness, 7
white matter, 25
wild type, 26
work study, 24
working memory, 10, 11, 14, 15, 18, 23, 36, 38, 39, 42

Y

yeast, 10
Y-maze, 15, 19, 23, 26